The Symbiotic Habit

The Symbiotic Habit

Angela E. Douglas

PRINCETON UNIVERSITY PRESS
PRINCETON AND OXFORD

Published by Princeton University Press, 41 William Street, Princeton, New Jersey 08540

In the United Kingdom: Princeton University Press, 6 Oxford Street, Woodstock,
Oxfordshire OX20 1TW

Library of Congress Cataloging-in-Publication Data

Douglas, A. E. (Angela Elizabeth), 1956–
The symbiotic habit / Angela E. Douglas.
p. cm.
Includes bibliographical references and index.
ISBN 978-0-691-11341-8 (hardcover : alk. paper) 1. Symbiosis. I. Title.
QH548.D678 2010
577.8'5—dc22 2009036344

British Library Cataloging-in-Publication Data is available

This book has been composed in Palatino

Printed on acid-free paper. ∞

press.princeton.edu

Printed in the United States of America

10 9 8 7 6 5 4 3 2 1

Contents

Preface

I HAVE BEEN INSPIRED to write this book by three developments that are revolutionizing the field of symbiosis.

The first development is technical: molecular and genomic techniques that enable us to identify essentially any symbiotic organism and to explain symbiotic function at the molecular level. The wealth of molecular data on symbiotic organisms accumulated over the last 10–15 years has transformed our understanding of the evolutionary origins and relationships of these organisms. There is now unambiguous evidence that symbiotic organisms have variously evolved from parasites, symbiotic partners in different associations, and organisms with no previous history of parasitism or symbiosis; and some symbioses are very ancient, while others have evolved over very short timescales, down to just hundreds of generations. In parallel, study of the ecology of symbioses has been changed radically by molecular tools that enable the abundance and distribution of genotypes to be quantified under field conditions. A topic that was previously dominated by studies of the ecology of organisms that happen to be symbiotic is increasingly addressing the very substantial roles of symbioses in shaping the structure of ecological communities, including their impact on the invasiveness of alien species. We are moving into the genomic age for symbiosis research, as illustrated by several published metagenomic analyses of microbial symbiont communities in animal hosts and the complete genomes of both partners available for at least two symbiosis (for the symbiosis between the legume *Medicago truncatula* and *Sinorhizobium* and the pea aphid–*Buchnera* symbiosis). We now have the opportunity to explain symbiosis function in genomic terms.

The second development is conceptual. It has long been recognized that symbioses are underpinned by the reciprocal exchange of benefit, and that the benefit can be costly to provide. This creates the potential for conflict between partners over the partitioning of resources. Symbiosis has traditionally been viewed as a balancing act in which each organism seeks to maximize its benefit, placing the association at perpetual risk of shifting to an exploitative relationship such as parasitism. It is increasingly realized that this perspective is inadequate: although partners cheat occasionally, symbioses rarely evolve into antagonistic relationships. Conflict in symbioses is managed effectively, generally by one partner taking control. There is now evidence that the controlling partner (generally the host) can operate in multiple ways. It can reward cooperating partners and impose sanctions against cheating partners, it can reduce conflict by controlling the transmission of its partners, and it can have specific recognition mechanisms that discriminate between acceptable and potentially deleterious partners. The concept of symbiosis as a mutually beneficial association in which conflict is managed by a controlling partner offers new insight into the processes underlying the exchange of benefits, and the establishment and persistence of stable symbioses.

The final development is the increasing application of symbiosis research to solve practical problems faced by humankind, as a direct result of the technical and conceptual advances described above. Undoubtedly, applied symbiosis research will only increase in importance as new opportunities emerge from our improved understanding of symbioses and also as the challenges of deleterious anthropogenic effects, such as habitat degradation and climate change, intensify. In particular, the devastation of some coral reef ecosystems by coral bleaching (caused by the breakdown of the coral-alga symbiosis), probably linked to climate change, has demonstrated the urgent need to understand the mechanisms underlying symbiosis persistence. We also have the potential of a rational basis to assess the feasibility of managing ecosystems for resistance to invasive symbiotic species or promoting symbioses for resistance to anthropogenic effects. Further opportunities include the potential to extend the range of plants capable of forming nitrogen-fixing nodules to major monocot crops, such as rice or wheat; and to enhance our own health and well-being by manipulating the composition of microorganisms in our digestive tracts.

These three developments have arisen since I wrote the books *Symbiotic Interactions* (1994) and *Biology of Symbiosis* (1987, coauthored with David Smith). The data were simply not available for detailed considerations of the evolutionary origins and fates of organisms in symbiosis or the consequences of conflict for symbiosis function. Similarly, the application of symbiosis to the ecological management, pest control, or medicine had barely started. For example, until recently, medical textbooks completely ignored the microbiota in our digestive tracts, even though it occupies a volume equivalent to the liver and is crucial to human immune function, resistance to pathogens, and nutrition, including propensity for obesity. This area is now a major area of symbiosis research.

This book has been a pleasure to write. It has been made possible by countless conversations and correspondence with colleagues who have introduced me to many remarkable symbiotic interactions and helped me to understand how symbioses work. I give particular thanks to Martin Bidartondo, Tom Boehm, David Clarke, Bryan Danforth, Martin Embley, Takeda Fukatsu, Ruth Gates, Toby Kiers, Margaret McFall-Ngai, Doyle McKey, Simon McQueen Mason, John Pringle, Rusty Rodriguez, Joel Sachs, Jan Sapp, David Schneider, Gavin Thomas, and Rachel Wood. I am grateful to David Smith for his enthusiasm and support of this project and to Barbara Brown, Jeremy Searle, and two anonymous referees whose comments on earlier drafts of the book have dramatically improved the text and corrected some factual errors. All remaining errors are of my making. I thank Alison Shakesby for assistance with the figures, and Sam Elworthy, Robert Kirk, and Alison Kalett, the three editors I have worked with at Princeton University Press, for their good advice and support. And finally, my thanks, as always, to Jeremy for his unfailing encouragement.

May your symbionts be with you.

Angela Douglas
1st January 2009

The Symbiotic Habit

CHAPTER 1

The Significance of Symbiosis

INDIVIDUALS OF DIFFERENT SPECIES form persistent associations from which they all benefit. These relationships are symbioses. The core purpose of this book is to assess the biological significance of symbioses and to investigate the processes by which symbioses are formed and persist in both evolutionary and ecological time. Symbioses are biologically important because they are widespread and dominate the biota of many habitats. I address this aspect of the biological significance of symbiosis later in this chapter. For the present, I suggest that any doubting readers should "look out of the window" and list every organism they can see. I can guarantee that most, probably all, organisms on the list are a product of symbiosis. The prevalence of symbioses is not, however, the only reason why symbioses should be important to biologists. An additional reason is that symbiosis challenges two widely accepted tenets of biology: the universality of descent with modification in evolution, and the primacy of antagonism in interactions among organisms. I will start by explaining these challenges.

1.1 Symbiosis as a Source of Novel Traits

The core expectations of evolution by descent with modification are that morphological, physiological, and other traits of an organism are derived from traits in the ancestors of the organism, and that changes in these traits can be described by multiple, small steps with each intermediate condition viable. Most traits can be explained in this way, but there is unambiguous evidence that some traits of great evolutionary and ecological importance have been gained laterally from different, often phylogenetically distant, taxa. Some laterally acquired traits are novel for the recipient organism and they can be evolutionary innovations, i.e., "newly acquired structures or properties which permit the assumption of a new function" (Mayr 1960, p. 351)].

There are two routes for the lateral acquisition of traits: symbiosis and horizontal gene transfer. Entire organisms are acquired by symbiosis and so the traits gained can be genetically, biochemically, and even behaviorally more complex than those obtained by horizontal transfer of isolated genes. Two very different types of symbiosis illustrate this point. The eukaryotic cells that acquired the cyanobacterial ancestor of plastids gained the capacity for oxygenic photosynthesis as a single package, including many correctly expressed genes, the integrated molecular and cellular machinery for the assembly of multicomponent photosystems in photosynthetic membranes, and the enzymatic machinery for carbon fixation. Similarly, an *Acacia* tree that is protected from herbivores by a resident colony of ants has acquired a morphologically and behaviorally complex defense capable of responding appropriately to herbivore attacks of variable magnitude and type.

The appreciation that descent with modification is an inadequate evolutionary explanation does not mean that symbioses contradict current understanding of evolutionary processes. Symbioses are subject to natural selection and, contrary to some claims (e.g., Ryan 2003), they have no discernible dynamic independent of natural selection.

1.2 Symbiosis as a Type of Biological Alliance

Interactions among organisms are routinely portrayed as principally antagonistic, dominated by competition, predation, and parasitism. The history of life has even been described as "a four billion year war" (Marjerus et al. 1996). This perspective is not wrong, but it is incomplete. Organisms have repeatedly responded to antagonists (predators, competitors, etc.) and abiotic stresses such as low nutrient availability by forming alliances, i.e., interactions with other organisms, resulting in enhanced fitness and ecological success of all the participants. As with alliances among people, political parties, and nation states, the persistence of many biological alliances depends on the continued presence of the antagonist, and the benefit gained from the alliance can vary with the identity of the participants and environmental conditions. I return to this issue in section 1.3.

Most alliances are founded on reciprocity, that it is advantageous to help another organism only if the favor is returned. In a two-organism system, reciprocity requires that each of the organisms places a higher value on what it receives (benefit) than what it gives (cost) (figure 1-1a). For example, the relationship between mycorrhizal fungi and plant roots is underpinned by the transfer of photosynthetic sugars from plant to fungus, and of phosphate in the reverse direction. Photosynthetic carbon is cheap for the plant to produce but a critical resource for the fungus, which cannot utilize the polymeric carbon sources in soil. Inorganic phosphate is relatively immobile in soils, and is acquired more readily by the fine, branching fungal hyphae than by the relatively

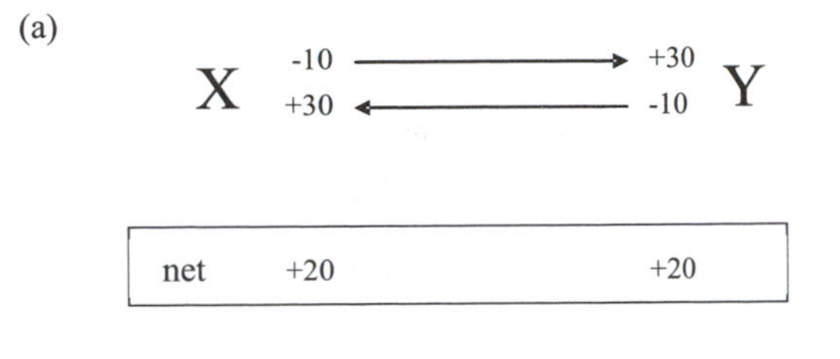

(b)

NUMBER OF SPECIES

One
>One
TRAIT
Same
Different
A
B
C
D

Figure 1-1 Biological alliances. (a) Most alliances are underpinned by reciprocity between the two participating organisms, X and Y. Each organism provides a service at cost of 10 arbitrary units and receives a benefit of 30 units, yielding net benefit of 20 units. (b) Alliances are classified according to the number of species and traits involved. Examples of each type of alliance include (A) roosting bats huddling together, sharing their uniform trait of heat production; (B) mixed-species flocks of passerine birds foraging for food in winter; (C) bartering of goods between humans; and (D) consortia of multiple microbial species that, through their complementary metabolic capabilities, degrade otherwise recalcitrant organic compounds. [Figure 1-1a modified from figure 1 of Douglas (2008)]

massive plant roots with short, nonbranching root hairs. The one situation where reciprocity as depicted in figure 1-1a does not apply is between closely related organisms. Here, kin selection is important: genotypes that help relatives (i.e., individuals with many genes in common) increase in frequency and are at a selective advantage over genotypes that do not help.

Many organisms in alliances display cooperative traits, i.e., traits that are advantageous to another organism (the recipient) and that have evolved because of their beneficial effect on the recipient. The *Acacia* trees introduced in section 1.1 display cooperative traits that benefit their resident ants: swollen, hollow thorns which provide domatia (nest sites) for the ants, and extrafloral nectar and highly nutritious antbodies on which the ants feed (figure 1-2). In return, various behavioral

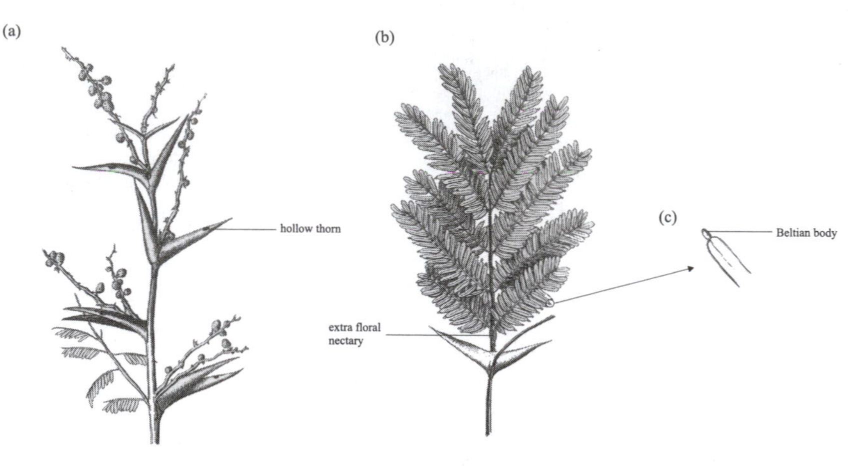

Figure 1-2 Adaptations of *Acacia sphaerocephala* for symbiosis with protective ants of the genus *Pseudomyrmex*. (a) Hollow thorns that provide domatia (nest sites) for ants. (b) Extrafloral nectary at base of leaf. (c) Lipid-rich ant-body at tip of leaflet. [Reproduced from figure 4-18 of E. O. Wilson (1971) *The Insect Societies*. Harvard University Press]

traits of the ants are advantageous to the plant, including patroling the plant to protect against herbivores and removing fungal spores at the breakpoint of the ant-bodies to prevent fungal infection of the plant. As a contrary example, the food in the gut of an animal infected by a tapeworm is not a cooperative trait because, although the tapeworm benefits from the food, animals have not evolved the habit of eating for the benefit of tapeworms.

Alliances can be classified according to whether they involve one or multiple species displaying the same or different traits, and this two-way classification generates four categories (figure 1-1b). The focus of this book is category D in figure 1-1b, alliances between different species with different traits. These alliances are also known as mutualisms, which are formally defined as relationships from which all participants derive benefit. In this book, I consider symbioses as a type of mutualism, specifically mutualisms in which the participants are in persistent contact.

This brings me to the most frustrating difficulty in the field of symbiosis—the lack of a single universally accepted definition. Disagreement over definitions has led to disputes about which relationships are symbioses and, consequently, a lack of consensus about the common features of symbiotic systems. Two alternative definitions of symbiosis, neither fully satisfactory, have dominated the literature for many decades: "symbiosis as any association" and "symbiosis as a persistent mutualism." Here, I digress briefly from the core topic of this chapter, the significance of symbioses, to address the thorny problem of definitions.

1.3 Definitions of Symbiosis

1.3.1 Symbiosis As Any Association

The term symbiosis was coined originally by Anton de Bary in 1879 to mean **any association between different species**, with the implication that the organisms are in persistent contact but that the relationship need not be advantageous to all the participants. De Bary explicitly included pathogenic and parasitic associations as examples of symbioses. Many symbiosis researchers use this definition and, without doubt, some colleagues steeped in the symbiosis literature will have objected to the opening two sentences of this book.

One key advantage of the definition of de Bary is that it promotes a broad context for research into symbioses. It acts as a reminder that it is important to investigate both the costs and the benefits to an organism of entering into a symbiosis (see figure 1-1a); and it is reasonable to expect some of the processes underlying relationships that are classified as mutualistic and antagonistic to be similar. For example, just as the persistence of certain antagonistic interactions depends on one organism failing to recognize the antagonist as a foreign organism, so some organisms may be accepted into symbioses because they fail to trigger the defense systems of their partner and not because they are positively recognized as mutualists.

Nevertheless, the definition of de Bary has two serious shortcomings. First and very importantly, the definition is not accepted by most general biologists or nonbiologists today, and so fails to communicate effectively. Most people do not describe the current malaria pandemic or the potato blight that caused the Irish famine of the 1840s as examples of symbiosis. Second, there are few principles generally applicable to symbioses, as defined by de Bary, but inapplicable to other biological systems. As a result, the "symbiosis as any association" definition is something of a catch-all category. Although this definition does promote further enquiry and insight into symbioses, any insights obtained are unlikely to be common to all symbioses defined in this way.

1.3.2 Symbiosis As a Persistent Mutualism

The definition of symbiosis widely accepted among both general biologists and the lexicographers who prepare English dictionaries is **an association between different species from which all participating**

organisms benefit. I subscribe to this definition even though it is not without difficulties.

The "symbiosis as a persistent mutualism" definition requires a formal assessment of the benefit derived by the organisms in the association. The standard approach to identify benefit is to compare an organism's performance (survival, growth, reproductive output, etc.) in the presence and absence of its partner. If the organism performs better with the partner, it benefits from the relationship and if it performs better in isolation, then it is harmed by the association. Although the methodology appears straightforward, it is unsuitable for many associations.

There are two types of problem. First, for some associations, there are formidable practical difficulties. Consider the deep-sea symbioses, such as the chemosynthetic bacteria in the tissues of pogonophoran worms at hydrothermal vents and the luminescent bacteria in the lure of deep-sea angler fish. It is difficult to envisage how bacteria-free pogonophorans and angler fish could be generated experimentally and how the performance of the bacteria-free individuals could be monitored reliably in habitats so inaccessible to humans.

The second and more fundamental problem is the variability of real associations, such that benefit is not a fixed trait of some relationships but varies, especially with environmental circumstance. To illustrate this issue, let us consider hermit crabs of the genus *Pagurus*. Hermit crabs live in empty shells of gastropods that are often colonized by benthic cnidarians, such as sea anemones and hydroids. Generally, the cnidarian benefits from settling onto a shell inhabited by a hermit crab because it has ready access to scraps of food produced when the crab eats and because the mobility of the crab in its shell introduces the cnidarian to different habitats that may increase food availability and reduce the risk of burial in sediment. The hermit crab is widely believed to benefit because the cnidarian can act as a bodyguard, protecting it from predators by firing deterrent and often toxic nematocysts from its tentacles. For example, the sea anemone *Calliactis parasitica* effectively deters octopus predation of *Pagurus* species (Ross 1971). The protective value of hydroids is, however, variable. Predation rates of hermit crabs can be elevated or depressed by hydroids, depending on the predator species, and the underlying factors are complex. For example, Buckley and Ebersole (1994) investigated *Pagurus longicarpus* inhabiting shells colonized by the hydroid *Hydractinia* spp. and subject to attack by the blue crab *Callinectes sapidus*. In aquarium trials, hermit crabs in shells bearing hydroids were significantly more likely than those in hydroid-free shells to be eaten by blue crabs (figure 1-3a). The difference arose because it took longer for the blue crabs to crush the hydroid-free shells, often giving the resident hermit crab time to escape. Further analysis of Buckley

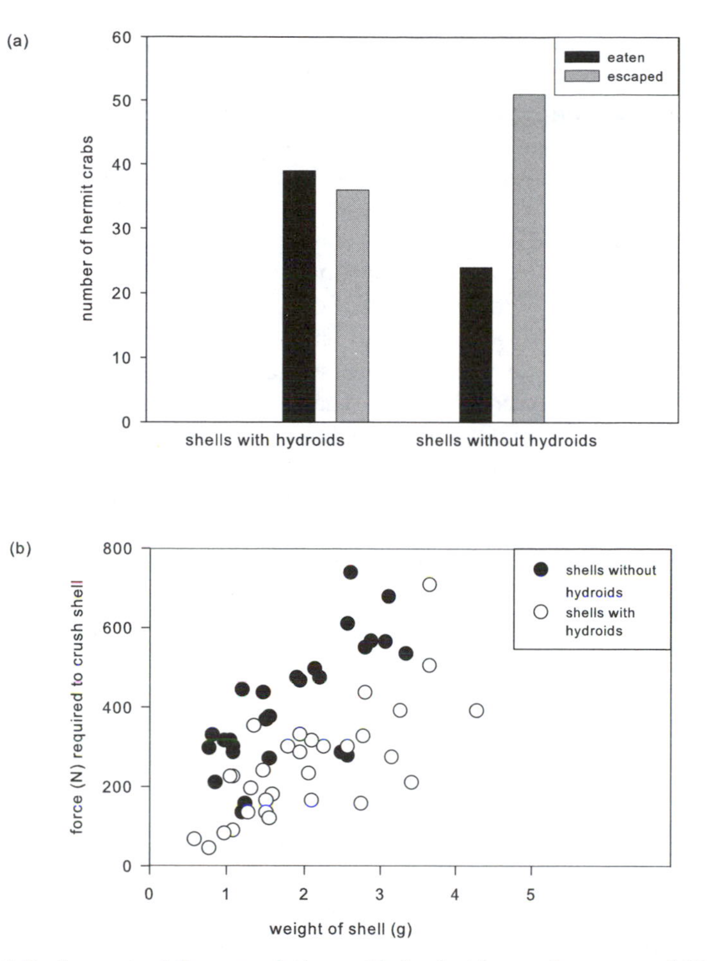

Figure 1-3 Impact of the association with hydroids on the susceptibility of the hermit crab *Pagurus longicarpus* to predation by *Callinectes sapidus*. (a) Occupation of hydroid-bearing shells significantly depressed the frequency of *P. longicarpus* that escaped from *C. sapidus* attack ($\chi^2 = 6.158$, $p < 0.05$). (b) Mechanical strength of shells inhabited by *P. longicarpus*, either colonized by hydroids (open symbols) or lacking hydroids (closed symbols). [Redrawn from data in Buckley and Ebersole (1994)]

and Ebersole (1994) revealed that shells bearing the hydroids were more likely than hydroid-free shells to be colonized by burrowing polychaete worms (*Polydora* species), and the tunnels of these worms significantly depressed the mechanical strength of the shells (figure 1-3b).

This complex multiway interaction between hermit crabs, hydroids, burrowing polychaete worms, and the predatory blue crabs raises a question: why do the hermit crabs ever use shells bearing the hydroids? One possible explanation is that the hydroids may protect the shell from colonization by large sessile animals, such as slipper limpets or bivalves, which would make the shell very heavy and unbalanced for the hermit crab. Based on these considerations, do hermit crabs benefit from associating with hydroids? The answer is that "it all depends"—on the incidence of burrowing worms, the identity and abundance of predators, and the incidence of settling limpets. All of these factors are anticipated to vary with site and season.

As the hermit crab association illustrates, many real associations are complex and variable. This does not undermine the definition of symbiosis as a mutually beneficial association, provided the definition refers to the interaction between the organisms, not the organisms themselves. An organism that enters into a mutually beneficial relationship is symbiotic in the context of that relationship; but if, through a change in environmental circumstance or other factors, the relationship becomes antagonistic, then it is no longer a symbiosis. In this way, associations with variable outcomes for the participating organisms are transformed from a problem for the definition of symbiosis to an opportunity to explore the factors that affect the incidence of symbiotic (i.e., mutually beneficial) interactions.

The fluidity of some biological interactions is particularly apparent for some organisms that were originally identified as parasites but are now realized to be harmless or even advantageous to partners under certain circumstances. For example, the fungus *Colletotrichum magna* was first identified as a virulent pathogen of certain plant species (figure 1-4a) but its impact on plant growth was subsequently found to depend on the plant species and even cultivar, such that the fungal infection promotes the growth of some plants (figure 1-4b) and can provide protection from drought or pathogens (Redman et al. 2001). Similarly, the bacterium *Helicobacter pylori* in the human stomach is best known as the cause of ulcers and gastric cancer in adults, especially older people, but in children and young people, *H. pylori* is harmless and may even be beneficial, providing protection against diarrhoea and asthma (Blaser and Atherton 2004). In the same way, organisms which are usually harmless or beneficial can be deleterious to their partners under certain circumstances. The tiny mite *Demodex folliculorum*, <0.5 mm long, is a commensal of humans. It lives exclusively in hair follicles, including the roots of eyelashes, and reproduces sexually with a generation time of 2–3 weeks. It occurs in most adults and many children, and it is generally harmless. Occasionally, however, the infections are heavy and deleterious, causing itching and swelling. Similarly, the arbuscular mycorrhizal fungi are usually beneficial symbionts of plant roots but they can,

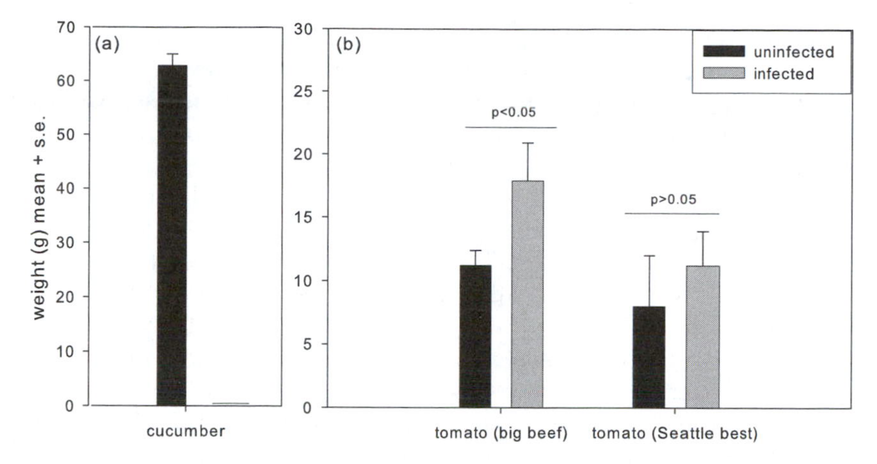

Figure 1-4 Impact of the fungus *Colletotrichum magna* (strain L2.5) on plant growth: (a) pathogenic on cucumber; (b) beneficial and apparently neutral for two tomato cultivars. s.e. indicates standard error. [Drawn from data in Redman et al. (2001)]

occasionally, be harmful to the plants. For example, *Glomus macrocarpum* causes a stunting disease of tobacco plants (Modjo and Hendrix 1986).

I favor the definition of symbiosis as a persistent mutualism because its scope provides a meaningful biological framework for one of the key questions explored in this book: How do mutually beneficial interactions between different species evolve and persist?

1.3.3 Duration of Contact

For how long should the partners in an interaction be in contact for the relationship to be called a symbiosis? It is widely, but not unanimously, accepted that the answer is: at least a substantial proportion of the lifespan of the interacting organisms. It has only rarely been argued that the predation of gazelles by lions, the rapid death of people infected by the Ebola virus, and the pollination of plants by fleeting visitations of foraging butterflies are symbioses (e.g., Lewis 1985).

Some examples bring into focus the problem posed by the definition of symbiosis as a persistent association. Let us consider first the relationship between humans and honeyguide birds, two species that share a taste for honey. The bird cannot access the honey from most bee colonies, but appears to have a better knowledge than people of the location of the colonies. In their study of the Boran people in Kenya, Isack and Reyer (1989) found that the honeyguide leads people to bee nests by calling and flying short distances ever closer to the nest, and then changes its call and flight pattern in the immediate vicinity of the nest. The people open up the nest to harvest the honey, and the bird feeds on the honeycomb fragments that are too small for the people to harvest. The participants in this relationship do not make contact and so the relationship is not a symbiosis.

At the other extreme of persistence are the vertically transmitted microorganisms in some animals, where contact persists beyond the lifespan of an individual host. The complement of intracellular bacteria in some insects is acquired exclusively from the mother insect, usually by transfer from the cells housing the bacteria to the ovaries and insertion into the unfertilized egg. Obligate vertical transmission can result in associations persisting continuously over long evolutionary timescales. As an example, there is good phylogenetic evidence that the bacterium *Buchnera aphidicola*, the vertically transmitted symbiont of aphids, has persisted through generations of its insect hosts for at least 160 million years (Moran et al. 1993); and the mitochondria have been transmitted vertically for up to 2000 million years (Embley and Martin 2006).

The difficulties for a definition of symbiosis as a persistent association arise for relationships with an intermediate duration of contact. This is illustrated by insect pollination of plants. For many plants, contact with the pollinator is brief, as little as a few seconds, while the insect collects nectar or pollen. As already considered, these interactions are not usually deemed to be symbioses. Among orchids, however, the duration of contact may be prolonged to minutes to hours, often with elaborate mechanisms to ensure pollen transfer to a specific location on the insect. This condition relates to the fact that the pollen grains of orchids are not separate but held in compact structures called pollinia, such that an orchid flower is dependent on a single insect for pollen transfer. An exceptional example is the orchid *Coryanthes*. As Barth (1991) describes, the complex *Coryanthes* flowers (figure 1-5) trap their pollinating bees, *Eulaema*, for long periods. The narrow base of the flower lip exudes a liquid scent, which the bee collects by scratching with its forelegs. It often slips off the lip, into the bucket below. The bucket contains a watery solution, exuded from the column of the flower, and the bee swims about in the water. The only exit is via a narrow hole at the apex of the lip, and it may take the bee half an hour or more to escape. If the bee is carrying pollinia from another flower, the pollinia are transferred to the stigma; and the pollinia from the current flower are transferred to the abdomen of the first bee to take this route through the flower. Even more persistent plant-pollinator contact is displayed by the plants with brood-site pollination, including the figs (*Ficus*) pollinated by agaonid wasps and *Yucca* species pollinated by moths. In these systems (discussed further in chapter 3, section 3.2.2), pollination is linked to insect

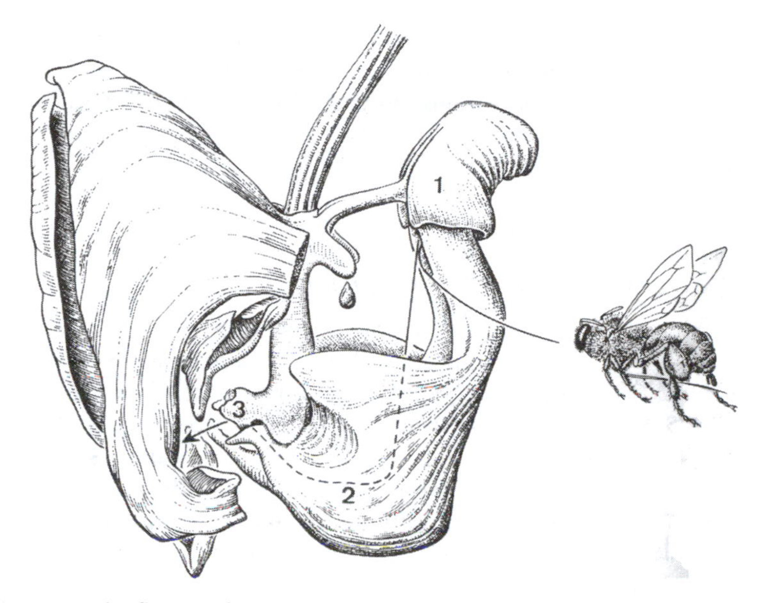

Figure 1-5 The flower of *Coryanthes* orchid. Arrows show the course of the pollinating insect. See text for details. [Redrawn from figure 7.27 of Barth (1991)]

deposition of an egg into some of the flower ovules. Seed set in the ovules bearing insect eggs is aborted and the insect offspring develops, to adulthood for the pollinators of *Ficus* and through larval development for the *Yucca* pollinators (which pupate in the soil).

A similar continuum in the duration of contact is displayed by the microbiota in animal guts. Members of the gut microbiota vary greatly in their persistence, from transients, entering and leaving the gut with the food, to residents, retained for much of the animal lifespan. A classic illustration is the study of *Escherichia coli* serotypes in a single human volunteer over 11 months. Of the 53 types identified, most persisted for one to a few days and just two types persisted for the full duration of the experiment (Caugant et al. 1981).

Among the *E. coli* types that vary in their persistence in the guts of humans and among the various pollination systems described above, there is no minimal residence time that can be used meaningfully as a criterion for symbiosis. In other words, it is biologically unrealistic to create a simple dichotomy based on duration of contact between relationships that are, and are not, symbioses. These issues have a bearing on the framework for this book. Although I will conform to the convention that relationships with no or fleeting physical contact are not symbioses, I will use these relationships as examples to illustrate and develop ideas that are relevant to symbioses.

Let us now return from the digression into definitions to the main purpose of this chapter: the significance of symbiosis. In the following section, I provide a brief primer on the evolutionary and ecological significance of some symbioses.

1.4 The Biological Significance of Symbioses

1.4.1 Misunderstandings about the Significance of Symbioses

The biological importance of symbioses is central to current understanding of these relationships. This opening sentence could not have been written throughout the greater part of the twentieth century, when mutually beneficial symbioses were treated as curiosities of nature that were ecologically unstable and evolutionarily transient (Sapp 1994). The ecological argument for the instability of symbioses was that, by Lotka-Volterra equations, mutually beneficial interactions lead to uncontrolled population increase while antagonistic interactions tend to stabilize populations (figure 1-6). The evolutionary argument was that genotypes which confer benefit on non-kin are at a selective disadvantage relative to selfish genotypes which provide no benefit. The implication is that symbioses fail in ecological time because they are too mutualistic and in evolutionary time because of the selection pressure to be less mutualistic. Both perspectives cannot be right and, in reality, both are wrong. The reasoning underlying each perspective is based on the erroneous assumption that symbioses are perfectly mutualistic. In reality, the partners in symbioses are often in conflict, but the conflict is managed and controlled. These issues are explored in chapter 3. Here, I consider the biological significance of symbioses in three respects: as a source of novel capabilities, in the evolution of eukaryotes, and as a determinant of the ecological success of some plants and animals.

1.4.2 Symbiosis As a Source of Novel Capabilities

For many symbioses, one organism provides a benefit that is different from any preexisting trait in its partner. As considered in section 1.1, such a benefit can be an evolutionary innovation. The novel capabilities gained in this way are very diverse but many can be classified as one of four functions: access to a novel metabolic capability, protection from antagonists (predators, pathogens, etc.), mobility (including dispersal), and agriculture. Table 1-1 provides some examples of symbioses and other mutualisms according to these functions.

For symbioses that involve reciprocal exchange of services, the benefits gained by the different partners can be functionally distinct or vary

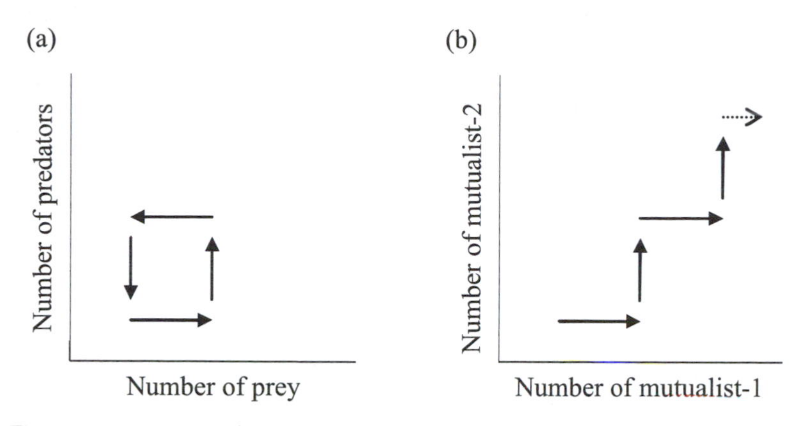

Figure 1-6 Projected population size of interacting organisms in (a) an antagonistic interaction such as a predator-prey relationship and (b) interaction between populations of two mutualists, here labeled as 1 and 2.

with circumstance. The *Acacia*-ant symbiosis illustrated in figure 1-2 involves the reciprocal exchange of food for protection as provided by the plant and ants, respectively. As considered earlier in this chapter (section 1.2), the mycorrhizal symbiosis between plant roots and fungi is primarily nutritional under most environmental conditions, with the plant gaining mineral nutrients from the fungus, and the fungus gaining photosynthetic sugars from the plant; but the interactions are not uniform. For example, the principal advantage of this symbiosis for natural populations of the plant *Vulpia ciliata* is protection from fungal pathogens (Newsham et al. 1995).

Animal behavior is central to the capabilities acquired by some symbiotic organisms. This applies particularly to animal-mediated dispersal, agriculture, and some instances of protection from predators (see table 1-1). Agriculture is of particular interest in this context. Agriculture can be defined as the cultivation of a different species and the consumption of a proportion of that organism or its products. It involves inoculating a suitable substratum with the cultivated organism, which is then tended to ensure its growth and protection from competitors and pathogens. Linked to the complex behavior required, farming has a very restricted distribution, largely limited to humans, ants, termites, and beetles. Fungi are farmed by attine ants, termites of the subfamily Macrotermitinae, and various beetles, including the ambrosia beetles (Platypodinae and some Scolytinae) (Wilding et al. 1989; Farrell et al. 2001). Some symbioses between ants and hemipteran insects also have the traits of agriculture, with the hemipterans farmed by ants for their honeydew (see chapter 5, section 5.2.1).

TABLE 1-1
Survey of Benefits Gained from Symbioses and Nonpersistent Mutualisms

Relationship	Examples
	(a) Access to metabolic capability
Inorganic carbon fixation	Cyanobacteria-derived plastids in algae and plants Algae/cyanobacteria in lichenized fungi, protists, and animals Mycorrhizal and endophytic fungi associated with plants Chemosynthetic bacteria in animals
Aerobic respiration	Bacteria-derived mitochondria in eukaryotes
Nitrogen fixation	Bacteria (e.g., rhizobia, *Frankia*, cyanobacteria) in plants Cyanobacteria in lichenized fungi Bacteria in a few insects (e.g., termites)
Cellulose degradation	Bacteria in vertebrates Protists in a few insects (e.g., "lower" termites, woodroaches) Fungi in Macrotermitidae
Nutrient biosynthesis (e.g., vitamins, essential amino acids)	Bacteria or fungi in animals, especially insects, and in protists
Degradation of toxins	Bacteria in animal guts
Toxin production	Bacteria in animals (e.g., insects, bryozoans, sponges) Fungi in plants
Hydrogen consumption	Methanogenic bacteria in anaerobic protists
Luminescence	Bacteria in some fish and squid
	(b) Protection from antagonists
Protection from herbivores	Ants associated with plants remove or deter herbivores
Protection from predators	Ants deter predators/parasitoids of hemipteran insects and lycaenids Sea anemone/hydroids protect hermit crabs
Removal of ectoparasites	From client fish by cleaner fish; from ungulates (e.g., gazelle) by pecking birds
Protection from pathogens	Microbiota in animal guts and plant rhizosphere (immediate environs of roots)

TABLE 1-1 (*cont.*)

Relationship	Examples
(c) Dispersal/mobility	
Biotic pollination	Male gamete (pollen) transported to stigma of plants by insects, birds, mammals
Biotic seed dispersal[1]	Seeds transported away from parent plant by birds, mammals, ants
Ant-tended hemipterans	Transport to suitable feeding sites on host plants
(d) Agriculture[2]	Cultivation of fungi by bark and ambrosia beetles, termites, and attinine ants Cultivation of hemipteran insects by some ants

[1]This category refers to seed dispersal by animals that forage for (a) seeds but fail to eat them all or (b) plant products associated with seeds, e.g., fruits consumed by various birds and mammals, elaiosomes (lipid-rich structures) consumed by ants. Seeds that become attached to the surface of animals (e.g., hooked seeds on the fur of mammals) are excluded because the interaction is passive and functionally comparable to wind dispersal (Herrera 2002).

[2]Defined as the cultivation of another organism and consumption of the cultivated organism or its products.

In many symbioses, the contrasting traits of the participating organisms are underpinned by a difference in their metabolic capabilities. The capabilities are complementary, with two or more organisms contributing different elements of a common metabolic pathway, resulting in the net synthesis or degradation of certain compounds. For example, bacteria of the genera *Clavibacter* and *Pseudomonas* can degrade the toxic herbicide atrazine to carbon dioxide and ammonia when cultured together but not individually (De Souza et al. 1998) because they have complementary metabolic capabilities (figure 1-7). These bacteria comprise a consortium, meaning that the interacting bacteria perform more complex capabilities than either displays on its own. In this particular system, the collective capabilities of the consortium are enhanced by cross-feeding of metabolites between the partners.

In other symbioses, the metabolic capabilities of the organisms are unequal, with one organism gaining access to capabilities present only in its partners. These associations generally involve eukaryotes, which, as a group, are metabolically impoverished relative to bacteria. The lineage that gave rise to the eukaryotes lacked the capacity for aerobic respiration, photosynthesis, and nitrogen fixation; many eukaryotes that

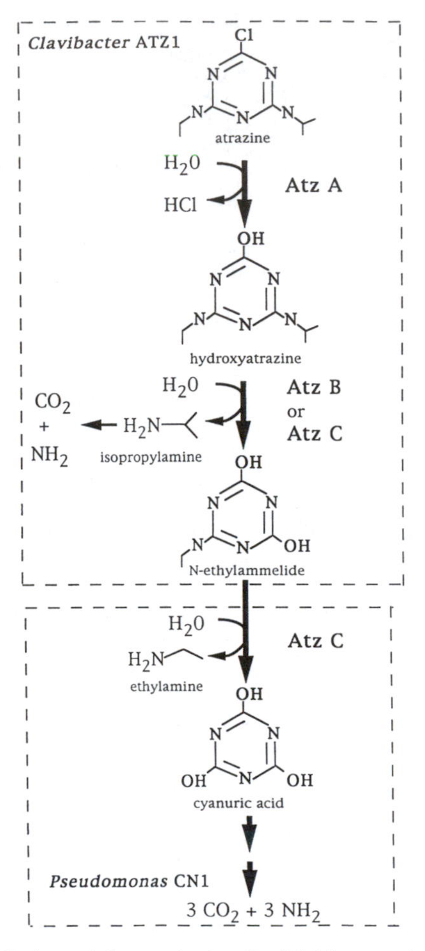

Figure 1-7 Degradation of the *s*-triazine herbicide atrazine to carbon dioxide and ammonia by a consortium of *Clavibacter* spp. and *Pseudomonas* spp. [Reproduced from figure 6 of de Souza et al. (1998) with permission from the American Society for Microbiology]

feed holozoically, including all animals, additionally cannot synthesize 9 of the 20 amino acids that make up protein (the essential amino acids) and various coenzymes essential to the function of enzymes central to metabolism (vitamins, such as biotin); vertebrates additionally lack the capacity to degrade the cellulose and related polysaccharides that account for 90% or more of plant material; and arthropods cannot synthesize sterols, including cholesterol which is an essential component of eukaryotic membranes. On multiple evolutionary occasions, eukaryotes have acquired these metabolic capabilities by entering into relationships with microorganisms. For example, various plants gain access to nitrogen fixation by associating with nitrogen-fixing bacteria; some animals, such as corals, and the lichen fungi gain photosynthesis by associating with algae; and other animals derive specific nutrients, including essential amino acids and B vitamins, from symbiotic bacteria. The services provided in these symbioses are nutritional.

Some organisms benefit from symbioses with partners which can synthesize secondary compounds that deter or are toxic to natural enemies. A well-known example among plants is the protection from herbivores gained by many grasses from alkaloid-producing endophytic fungi in their tissues (Schardl 1996; see also chapter 2, section 2.2.3). The incidence of these symbioses between animals and microorganisms is uncertain, and only a few examples have been explored in detail. They include the production of compounds known as bryostatins by the bacterium *Endobugula sertula* in the marine bryozoan *Bugula neritina* (Davidson et al. 2001) and the polyketide pederin by uncultured pseudomonads in both beetles of the genus *Pederus* and sponges, including *Theonella swinhoei* (Piel 2002; Piel et al. 2004). There is evidence, direct for *Pederus* and circumstantial for the bryozoans and sponges, that these compounds protect the animals from predators (Kellner and Dettner 1996; McGovern and Hellberg 2003).

The complexity of some symbioses involving microorganisms with a defensive function is illustrated by two highly specific associations between insects and streptomycete bacteria. One is the relationship between beewolves (bees of the family Crabanidae) and Candidatus *Streptomyces philantii* located in glands on the antennae of the adult beewolf (Kaltenpoth et al. 2005). The adult female transfers an inoculum of bacteria to the brood cell, in which eggs are deposited. In due course, the larval offspring collect the bacteria from the brood cell and apply them to the silk of their cocoon within which they subsequently pupate in the soil. Behavioral and bioassay data indicate that the streptomycete produces antibiotics which protect the pupa from fungal attack. Streptomycetes are also crucial for leaf-cutting ants. These ants maintain fungal gardens of a specific basidiomycete fungus on which the ants feed. The fungus is susceptible to a specific fungal pathogen *Escovopsis*, and *Escovopsis* infections are kept in check by antibiotics produced by streptomycete associated with the ventral surface of the ants (Currie et al. 2003).

Bioactive compounds synthesized by a symbiotic partner can also be exploited in offense, as is illustrated by antlions, predatory larvae of the insect family Myrmeleontidae within the order Neuroptera (the lacewings). Antlions subdue their prey by injecting toxic saliva via their piercing mouthparts. The salivary toxin which paralyzes the prey of *Myrmeleon bore* larvae is a protein, GroEL, produced by the bacterial symbiont *Enterobacter aerogenes* (Yoshida et al. 2001). GroEL is a well-known bacterial protein that functions as a chaperonin, i.e. it promotes the correct folding of proteins. It appears that the GroEL of *E. aerogenes* has adopted a second function as a toxin while retaining its chaperonin function. Four amino acid residues have been identified as critical to the toxicity of the protein; they are absent from the GroEL of other bacteria, including *E. coli* (the *E. coli* GroEL is not toxic). The mode of action of the GroEL from the antlion symbiont is not known, but it appears to be relatively specific causing paralysis of other insects, including cockroaches, but not of mice. How the antlion avoids the toxic effects of the protein is also unknown.

Some microrganisms protect plants and animals from toxic organic compounds and metals in the environment. One of the most remarkable examples relates to the potent neurotoxin methylmercury, the pollutant which caused large-scale poisoning of the human population of Minamata, Japan in the 1950s. Methylmercury is absorbed very efficiently across the gut wall of animals because it forms a complex with the amino acid cysteine that mimics another amino acid, methionine, resulting in its transport throughout the body (Clarkson et al. 2003). Bacteria in the gut of mammals can degrade methylmercury by demethylating it to elemental mercury and mercuric ions, both of which are eliminated via the feces (Rowland 1999). This reaction in the symbiotic bacteria has provided people with a degree of protection against methylmercury contamination of foods.

1.4.3 *Symbiosis and the Biology of Eukaryotes*

The eukaryotic condition is fundamentally symbiotic. The basis of this statement is that all eukaryotes either bear mitochondria or are derived from mitochondriate ancestors; and mitochondria have evolved from symbiotic bacteria. The evidence for a symbiotic origin of mitochondria is overwhelming. Mitochondria are produced exclusively by division of preexisting mitochondria, i.e., eukaryotic cells cannot generate mitochondria *de novo*, and virtually all mitochondria have their own genome including genes of sequence allied to those of α-proteobacteria, specifically rickettsias (Gray et al. 1999). Certain anaerobic protists including the diplomonads, metamonads, and microsporidians have proven to be crucial to our understanding of the centrality of symbiosis to eukaryotes. These protists bear no structures readily identifiable with mitochondria but, where studied, they have genes of sequence allied unambiguously to mitochondrion-derived genes. The interpretation that these organisms have mitochondriate ancestors is supported by careful cytological analysis revealing tiny membrane-bound organelles, known as mitosomes, with the characteristics of relict mitochondria (e.g., Tovar et al. 2003).

The eukaryote-mitochondrial relationship probably had a single evolutionary origin at least 1.45 billion and perhaps 2 billion years ago (Embley and Martin 2006), meaning that all modern eukaryotes have evolved in the context of a long-standing intimate symbiosis with a foreign organism in their cytoplasm. The implications are considerable, as can be illustrated by programmed cell death in eukaryotes. An essential early step in the commitment of eukaryotic cells to die is signal exchange between the cytoplasm and the mitochondrion, resulting in the release of the mitochondrial protein cytochrome *c* from the mitochondrial matrix to the cytoplasm. Why should an organelle with primary function in aerobic metabolism be central to cell suicide? One possibility is that programmed cell death is an evolutionary modification of antagonistic interactions between the eukaryotic nucleocytoplasm and the bacterial ancestor of mitochondria. It is not suggested that programmed cell death is an expression of virulence in modern mitochondria, but that the central role of mitochondria in programmed cell death may have evolved from such an interaction. I consider this interaction further in the context of the evolution of mitochondria in chapter 3, section 3.6.3.

It is widely accepted that the selective advantage to the eukaryote of associating with the bacterial ancestor of mitochondria was access to the bacterial trait of aerobic respiration. The ancestral eukaryotes also apparently could not fix inorganic carbon from CO_2 or nitrogen from elemental nitrogen, N_2. Various eukaryotic lineages have repeatedly acquired these latter capabilities by symbiosis with bacteria possessing these capabilities. The plastids/chloroplasts of all photosynthetic eukaryotes (the plants and algae) are derived ultimately from symbiotic cyanobacteria capable of oxygenic photosynthesis. There are various symbioses between eukaryotes and nitrogen-fixing bacteria (Kneip et al. 2007), but we have no entirely satisfactory explanation for the apparent absence of nitrogen-fixing organelles in eukaryotes (see chapter 3, section 3.6.5).

1.4.4 Symbiosis and the Ecological Significance of Some Plants and Animals

Plants and animals live in a microbial world. Their surfaces are colonized by microorganisms (bacteria and protists) from which they generally derive no substantial harm. Some plants and animals, however, live in specific and coevolved relationships with particular microorganisms, and these associations have profound impacts on the ecology and evolution of the taxa involved and, in some instances, also on entire ecosystems. In particular, animal or plant symbioses with microorganisms dominate most terrestrial landscapes, certain coastal environments and the immediate environs of deep-sea hydrothermal vents.

The roots of more than 75% of plant taxa are susceptible to infection by mycorrhizal fungi (Newman and Reddell 1987) which generally promote plant mineral nutrition. Molecular and paleontological data suggest that the association with one type of mycorrhizal fungi, the arbuscular mycorrhizal fungi, is very ancient, probably evolving ca. 400 million years ago at the time of the origin of land plants (Simon et al. 1993). It has been argued that the symbiosis and the resultant enhanced capacity of early plants to acquire nutrients from the substratum was a prerequisite for the evolutionary transition of plants from aquatic to terrestrial habitats. Furthermore, these fungi produce a glycoprotein known as glomalin which can account for up to 5% of soil carbon and nitrogen and promotes soil structure (Rillig 2004; Treseder and Turner 2007). The implication is that, if mycorrhizal associations had not evolved, then terrestrial landscapes would probably have been dominated by microbial mats and crusts, especially of cyanobacteria. Certainly, mycorrhizal infection is the norm in the vegetation in most habitats today, including tropical rainforests, temperate and boreal forests, savannah, and temperate grasslands (Treseder and Cross 2006). One important type of vegetation that is not founded on mycorrhizas is land used for annual crop production, where plowing and related practices generate a highly disturbed soil environment in which mycorrhizal fungal networks cannot establish fully. This is one factor contributing to the requirement of conventional crop production for high nutrient inputs.

Wave-resistant calcareous reefs are a characteristic feature of many shallow, clear waters at low latitudes (<ca. 35°). Although these reefs account for just 2% of the area of coastal waters, they are of immense ecological and economic importance. They are renowned as ecosystems of high biodiversity and productivity, and they support fishing and tourism industries crucial to the economy of many countries, as well as providing coastal defense against storms (Cesar et al. 2003). The reefs are generated predominantly from the skeletons of scleractinian corals bearing dinoflagellate microalgae of the genus *Symbiodinium* (also known as zooxanthellae) in their tissues; and there is strong experimental evidence that coral skeletogenesis by shallow-water corals is promoted by the photosynthetic activity of the zooxanthellae (Moya et al. 2006). The coral-zooxanthella symbiosis is, thus, the architectural foundation of shallow-water coral reef ecosystems. Furthermore, there is evidence that reef building by scleractinian corals in shallow waters

depended on the prior evolution of the algal symbiosis. This issue is considered further in chapter 5 (section 5.2.1).

The importance of microbial symbioses in shaping ecological communities is also illustrated by the fauna associated with deep-sea hydrothermal vents, where hot waters enriched with reduced inorganic compounds are released into the water column at areas of sea floor spreading. One would expect such habitats to be dominated by microorganisms, especially chemoautotrophic bacteria which can harness the energy generated by oxidation of reduced inorganic compounds to carbon dioxide fixation. Most hydrothermal vents, however, are also richly colonized by a diverse invertebrate fauna, in some locations including vestimentiferan tube-worms up to 2 m long and large bivalve mollusks. The basis of this exceptional fauna is symbiosis: the animals derive fixed carbon compounds from chemoautotrophic bacteria in their tissues (Van Dover 2000). These communities are unique in their independence from solar energy. They could persist for millenia in the absence of the sun (but not indefinitely because they require oxygen, derived ultimately from oxygenic photosynthesis). Taxonomically allied worms and bivalves also occur in marine sediments, hydrocarbon seeps, and whalefalls, i.e., the decaying carcasses of whales on the sea bottom (Van Dover 2000; Baco and Smith 2003). Symbiosis with chemoautotrophs has expanded the metabolic repertoire of these animals, enabling them to exploit habitats that would otherwise be available only to microorganisms.

In addition to the many symbioses with microorganisms, a wide diversity of animals and plants associate with other animals, e.g., mites and ants with plants; shrimps, crabs, and fish with corals and sea anemones. These symbioses have traditionally been treated as ecologically trivial but useful model systems to study the behavioral and morphological consequences of coevolutionary processes and biological specialization. Recent research on both some marine relationships and ant associations with plants and hemipteran insects suggests that these associations are of far greater ecological significance than traditionally believed. In particular, some have substantial impacts on the wider community structure, including the promotion of species diversity (e.g., Heil and McKey 2003; Hay et al. 2004).

1.5 The Structure of This Book

In this chapter, I have introduced and illustrated two key points. First, we live in a symbiosis-rich world, in which the symbiotic habit is widespread and abundant. Second, symbioses contribute to the real complexity and variability in biotic interactions. In particular, the prevalence of symbioses demonstrates that the natural world is not driven exclusively by antagonistic interactions; and the acquisition of novel traits by symbiosis is an important exception to the generality that evolutionary change is mediated by multiple, sequential mutations, each with a small phenotypic effect. Altogether, symbiosis is a biological phenomenon of first-order importance.

As stated at the beginning of this chapter, the central purpose of this book is to explore the symbiotic habit, including how symbioses are formed and persist in both evolutionary and physiological time. Symbioses pose genuine biological problems. As the many twentieth-century biologists who regarded symbioses as trivial would argue, symbiosis appears to be improbable because the benefits that symbiotic organisms confer on their partners are often costly to provide, and the participants often compete for a common resource. How is it that, despite these conflicts, symbioses are prevalent and persistent? The answer comes in three parts, explored in chapters 2–4.

In chapter 2, "The Evolutionary Origins and Fates of Organisms in Symbiosis," I consider the evolutionary relationships between symbiotic organisms and organisms with different lifestyles. It is widely assumed, especially among biologists modeling the evolution of cooperation and symbiosis (e.g., Johnstone and Bshary, 2002; Hauert et al., 2006; Traulsen and Novak 2006; Taylor et al. 2007) that symbioses have evolved from antagonistic relationships and are prone to revert to antagonism. The review of the evolutionary history of symbioses in chapter 2 yields a very different picture. Although evolutionary transitions between antagonistic and mutualistic relationships certainly have occurred, the evolutionary origins of symbioses are diverse, and many real symbioses are derived from casual relationships or involve mutualists derived from different symbioses.

Whatever the evolutionary origin of symbioses, their persistence implies that conflicts among the partners do not necessarily lead to symbiosis breakdown within the lifespan of an individual organism or over short evolutionary timescales. In chapter 3,"Conflict and Conflict Resolution," I review the incidence of conflict and some of the routes by which conflict can be managed. Potentially widespread mechanisms of conflict resolution involve asymmetry among the symbiotic organisms, with one in control and enforcing the good behavior of its partners. Although symbioses are mutualistic, they are not necessarily cooperatives of coequals. This is illustrated particularly by organelles that are controlled almost totally by the host. I argue that the evolutionary transition to organelles is the consequence of one route to conflict resolution.

Conflict among organisms in symbiosis is heightened when one partner fails to provide a service or consumes common resources excessively. Symbiotic organisms that avoid such unsuitable partners are expected to encounter lower levels of conflict with their partners. Partner choice is a sufficiently important topic to merit a full chapter on its own. In chapter 4, "Choosing and Chosen in Symbiosis," I explore the mechanisms by which organisms identify cooperative partners, discriminate against ineffective organisms and persist together in apparent harmony. Many of the putative mechanisms are widely perceived as preexisting defenses against antagonists that are recruited, often with modification, for symbiotic function. However, with the recognition of the antiquity and pervasiveness of the symbiotic habit, we should consider alternative possibilities. Perhaps, mechanisms that have evolved for the management of organismal interactions can be applied to both beneficial and antagonistic relationships; and they are labeled as defensive responses only because of the greater research effort on antagonistic than on mutualistic associations.

In chapter 5, "The Success of Symbiosis," I apply understanding of the formation and persistence of symbioses, as explored in chapters 2–4, to investigate the basis of the significance of the symbiotic habit. Symbioses are important not just because they are widespread and abundant (as considered in this chapter) but also because the acquisition of symbiosis can dramatically alter the evolutionary history of some lineages and change the structure of ecological communities. I also consider the success of symbioses in the context of human activities. There are both threats and opportunities: threats to symbioses arising from anthropogenic impacts, and opportunities where we can harness symbioses to our advantage, especially in medicine and pest control strategies. Understanding of the processes underlying the symbiotic habit (chapters 2–4) can contribute to the amelioration of these threats and the successful exploitation of various associations.

In the concluding chapter 6, "Perspectives," I explore the opportunity for future research on symbiosis and some outstanding questions that can be resolved best by analysis across many symbioses.

The structure of this book is founded on the perspective that a grasp of multiple systems can promote our understanding of the symbiotic habit. Most of the systems that I use to illustrate and develop themes are symbioses, but I also address parasitic or pathogenic interactions and nonpersistent mutualistic relationships where they can contribute to my argument.

CHAPTER 2

Evolutionary Origins and Fates of Organisms in Symbiosis

THE MAIN CONCLUSION from chapter 1 is that the symbiotic habit is a significant source of evolutionary innovation and is ecologically important. In this chapter, I focus on the organisms contributing to symbioses, specifically to address their evolutionary history. How frequently have organisms evolved the capacity to form symbioses, and are organisms of a particular phylogenetic position or lifestyle especially predisposed to evolve this trait? Do symbioses regularly evolve and break down, and how common are the evolutionary transitions between antagonistic and mutually beneficial relationships?

To a large extent, the answers to these questions lie in the phylogenetic distribution of symbiotic organisms. Some symbioses have an accessible fossil history. For example, there is paleontological evidence for lichens 600 million years ago (Yuan et al. 2005) and arbuscular mycorrhizal fungi 400 million years ago (Taylor et al. 1995). It has also become increasingly possible to explore the evolutionary history of organisms that leave no trace in the fossil record and are not amenable to comparative morphological analysis. The availability of molecular techniques and sophisticated computational tools for phylogenetic reconstruction makes rigorous evolutionary analysis of essentially any organism feasible. These tools are being applied to a variety of symbioses. A word of caution is, however, needed because some molecular phylogenies, especially those based on single genes, with uneven representation of sequences from different taxonomic groups or including anciently diverged sequences, might be incomplete and even erroneous. For many symbioses, current evolutionary understanding is more work in progress than the last word.

A reasonable starting point is that symbiosis is a derived state: that all organisms in symbiosis have ancestors that were not symbiotic. Nevertheless, many symbiotic organisms have a long and often complex evolutionary history in symbiosis. Much of the research on the evolutionary history of symbiotic organisms has concerned two types of event: the transitions between antagonistic and mutually beneficial relationships, and partner capture, i.e., the acquisition of novel partners by organisms with a prior history of symbiosis. I will consider these two topics in turn (sections 2.2 and 2.3) before addressing the evolutionary synthesis of symbioses between organisms with no immediate symbiotic history and the reverse process of the evolutionary loss of the symbiotic habit (section 2.4).

2.2 Evolutionary Transitions Between Antagonistic and Symbiotic Relationships

2.2.1 *Organisms with Variable Impacts on their Partners*

As considered in chapter 1, certain organisms can be beneficial or deleterious to their partners, varying with environmental circumstance (chapter 1; see section 1.3.2). We have some understanding of the mechanisms underlying the capacity of a few of these remarkable organisms to switch between antagonistic and symbiotic lifestyles.

Bacteria of the genus *Photorhabdus* combine two lifestyles, as virulent pathogens of insects and as symbionts required for growth of heterorhabditid nematode worms. These two lifestyles are tightly choreographed to occur at different stages in the lifecycle of the nematode. The nematode is a parasite of insects living in soil. *Photorhabdus* has two habitats: in the gut of the free-living nematode, where it is a symbiont required for nematode growth and persistence; and expelled from the nematode into the insect host of the nematode, where it is a pathogen. Within the insect, the bacteria multiply rapidly, supported by insect-derived nutrients, and produce toxins and antibiotics that prevent invasion by other organisms. The nematodes also feed on the bacteria. When conditions in the cadaver deteriorate through crowding and nutrient depletion, the nematodes produce a nonfeeding juvenile form, known as the infective juvenile stage, which is colonized by some of the remaining *Photorhabdus* in the insect and then emerges into the soil to seek a new insect host.

Overlapping sets of *Photorhabdus* genes required for symbiosis and pathogenesis have been identified, with a greater number required for symbiosis than for pathogenesis. Furthermore, these various genes are regulated by specific genes. One, known as *hexA*, is a negative regulator of symbiosis and positive regulator of pathogenesis; while *phoPQ* is absolutely required for pathogenicity and is a positive regulator of some symbiosis genes (Joyce et al. 2006) (figure 2-1). Various bacteria (e.g., *Erwinia*, *Salmonella*) related to *Photorhabdus* possess homologs of *hexA* and *phoPQ* that function as regulators of pathogenicity, raising the possibility that genetic circuitry which evolved in strictly pathogenic contexts has subsequently been modified to control both the symbiotic and pathogenic interactions in *Photorhabdus*. More generally, the

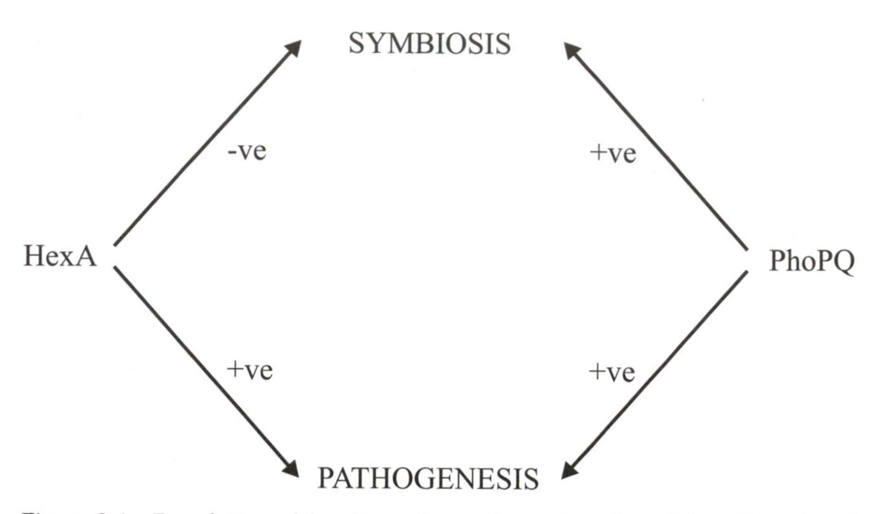

Figure 2-1 Regulation of the alternative pathogenic and symbiotic lifestyles of the bacterium *Photorhabdus* by the gene products HexA and PhoPQ. [Redrawn from figure 2 of Joyce et al. (2006)]

Photorhabdus system illustrates how pathogenesis and symbiosis can be underpinned by common molecular mechanisms.

For some taxa, the organisms displaying pathogenic and symbiosis lifestyles are not genetically identical but differ by just one or a few genes. This has been revealed by an elegant study of an endophytic fungus that lives in the shoots of grasses. The wild-type fungus *Neotyphodium* in the perennial rye grass *Lolium perenne* is not deleterious (it either has no impact on plant growth or is beneficial, depending on the environmental conditions). A mutant screen of the fungus identified a single gene, *noxA,* coding NADPH oxidase that is crucially important to the relationship (Tanaka et al. 2006a). When *noxA* was inactivated, the fungus became detrimental to the grass. The fungal biomass in the shoots increased dramatically, and the plant grew slowly and died prematurely. The basis of this effect is uncertain but reactive oxygen species generated by NADPH oxidase in the wild-type fungus may activate fungal signaling pathways that regulate fungal growth (see chapter 4, section 4.5.1). The important point in the present context is that the function of a single gene defines whether the fungus is a symbiont or pathogen.

There is also excellent evidence that the presence or absence of a small number of virulence genes can dictate whether certain bacteria are pathogenic or not. In many cases, the virulence genes are horizontally transmissible, borne on plasmids, bacteriophage, or pathogenicity islands integrated into the bacterial chromosome. (Pathogenicity islands are chromosomal regions flanked by insertion sequences or direct repeats and many have a different G + C content from the rest of the genome.) For example, the gain or loss of specific pathogenicity islands coding for adhesive fimbriae, iron-uptake systems, and hemolysin can dictate whether *Escherichia coli* cells live harmlessly in our digestive tract or cause a pathological infection of the urinary tract (Pallen and Wren 2007).

The lateral transfer of genes has also been implicated in the evolutionary transition between pathogenesis and symbiosis in certain bacteria associated with plant roots. The bacteria are *Rhizobium*, a nitrogen-fixing symbiont in root nodules of leguminous plants, and *Agrobacterium*, which induces pathological growth disturbances in plant roots (e.g., tumors, excessive root proliferation) and associated stunting of plant growth. Although the symbionts and pathogens are allocated to different genera, some species of *Rhizobium* are more closely related to *Agrobacterium* than to other *Rhizobium* species (Young et al. 2001). In both *Rhizobium* and *Agrobacterium*, the genes conferring the capacity to infect plants are plasmid-borne and horizontally transmissible, in principle allowing for bacteria of a single chromosomal genotype to adopt either lifestyle as dictated by their plasmid complement. For example, *A. tumefaciens* manipulated to bear the relevant plasmid (known as the Sym plasmid) from *R. etli* makes nitrogen-fixing nodules on *Phaseolus* bean plants (Martinez et al. 1987).

The existence of organisms which can switch between pathogenic and symbiotic lifestyles through changes in gene expression, mutation, or horizontal gene transfer indicates that the evolutionary transition between predominantly or exclusively antagonistic and symbiotic lifestyles is feasible. Let us now consider the incidence and processes involved in the evolutionary transitions, first from antagonistic interactions to symbioses and then in the reverse direction.

2.2.2 *From Antagonism to Mutualism*

Plausible scenarios can readily be constructed for the evolutionary transition from antagonistic to mutualistic relationships among non-symbiotic systems (i.e., those that do not involve persistent contact between the partners). Pollen and seed dispersal by feeding animals can reasonably be inferred to have evolved from animal predation of pollen and seeds. The shift can be attributed to plant-mediated manipulation of animal behavior, such that the advantage to the plant of the animal trait of mobility is enhanced, while the negative consequences of a second trait, food consumption, are reduced. Plants have manipulated the relationship in their favor in several ways. The sculptured pollen surface, mediated by irregularities in deposition of the pollen wall component sporopollenin, promotes adherence of pollen to the pollinator; seed production can be synchronized, so that predators are satiated and do not eat all the seeds that they harvest and disperse; and many plants reduce pollen or seed consumption by providing an alternative food source (nectar or fruit) that is cheap to produce and attractive to the animal. Nectar is provided in small quantities, so that the animal is not satiated rapidly and forages for long periods, promoting pollination.

Ant-tending of insects might also have evolved from a predator-prey relationship. Ancestrally, ants are generalist predators; and the protection of tended insects from predators is simply protecting a source (the insect) of sugars. The initial stages in the evolutionary transition from predation to tending might have been facilitated where the tended insect is nutritionally of low quality and not the preferred prey of ants; this applies particularly to aphids (Stadler and Dixon 2005). Some aspects of the relationship support this scenario. In particular, ants consume small numbers of the tended insects when the tended insects are very abundant or when alternative sources of food are not available (Way 1963; Offenberg 2001). The transition to tending may have been facilitated by the preexisting behavioral repertoire of the ants. Specifically, ants solicit honeydew from hemipterans in essentially the same way as they induce nest mates to regurgitate food.

There are firm phylogenetic data supporting the parasitic origin of some symbiotic microorganisms. This particularly applies to fungi of the family Clavicipitaceae, including *Epichloë* and *Cordiceps*. *Epichloë* parasites of grasses have given rise to beneficial endophytic fungi, as described in the following section (section 2.2.3). Most *Cordiceps* species are pathogens of insects or less commonly fungi, but this genus also includes symbionts required by delphacid plant-hoppers (e.g., the brown rice plant-hopper *Nilaparvata lugens*) and cerataphinid aphids (e.g., *Hamiltonaphis styraci*) (Suh et al. 2001). The symbionts are traditionally known as yeastlike symbionts because they live as single cells in the body cavity and fat body of their insect hosts, unlike the pathogenic *Cordiceps* which are filamentous and ramify throughout the insect tissues. The pathogen-derived symbionts in the plant-hoppers and aphids are sister taxa that recycle nitrogen and provide sterols to their hosts (Hongoh and Ishikawa 2000; Noda and Koizumi 2003).

What are the processes underlying the transition from antagonist to mutualist? Two processes that have been invoked: amelioration and addiction, and they are considered in the following sections.

2.2.3 *Amelioration*

Amelioration is the reduction of virulence. This process is strongly favored in parts of the parasitology literature. The virulence of a parasite is predicted to decline with time because the parasite derives no advantage from debilitating the host, and virulent parasites are deemed to be poorly adapted. For example, a parasite can be virulent because it is in the wrong tissue (e.g., *Neisseria meningitidis* only causes meningitis when it colonizes the cerebrospinal fluid; in the nasopharynx, it is asymptomatic) or inadvertently triggers an excessive immune response in the host (e.g., the damage to the peripheral nervous system in leprosy is caused by the human immune response to *Mycobacterium leprae*). Such effects are of no advantage to the parasite and are predicted to be selected against, resulting in reduced virulence to the host.

Nevertheless, amelioration alone cannot explain the evolutionary transition from antagonism to mutual benefit for two reasons. The first is that the selection pressure on a parasite is to maximize its reproductive output, not to display reduced virulence. Amelioration is predicted to occur only if it increases parasite fitness. A major determinant of the relationship between parasite fitness and virulence is mode of transmission. If transmission is promoted by keeping a host alive or healthy, less virulent parasites are at a selective advantage. In broad terms, the patterns of parasite virulence illustrate this generality (Ewald 1994). Thus, bacterial pathogens transmitted by contagion tend to be less virulent than those transmitted by water; parasites transmitted by insect vectors tend to debilitate the host but not the vector because transmission is promoted by a mobile, active insect and morbid host too sick to deter biting insects; and taxa that are transmitted exclusively vertically tend to be less virulent than taxa with horizontal transmission because the success of vertical transmission is correlated with the number and vigor of host offspring.

The second limitation of amelioration as an explanation for the evolution of mutualisms is that it generates a less virulent parasite, not a symbiont. This can be illustrated by an unplanned but most informative evolutionary experiment involving viruses: the introduction of myxoma virus to populations of the European rabbit *Oryctolagus cuniculus* in Australia in 1950. The myxoma virus occurs naturally in the South American cotton-tail rabbit *Sylvilagus brasiliensis*, in which it causes mild disease. When initially released into the rabbit populations in Australia, the virus was highly virulent, killing most host individuals in two weeks (Fenner 1983). However, within a few years, the virulence of the virus declined to an intermediate level and, a decade later, the most common virus genotypes remained of intermediate virulence (figure 2-2). The amelioration of virulence could be linked to

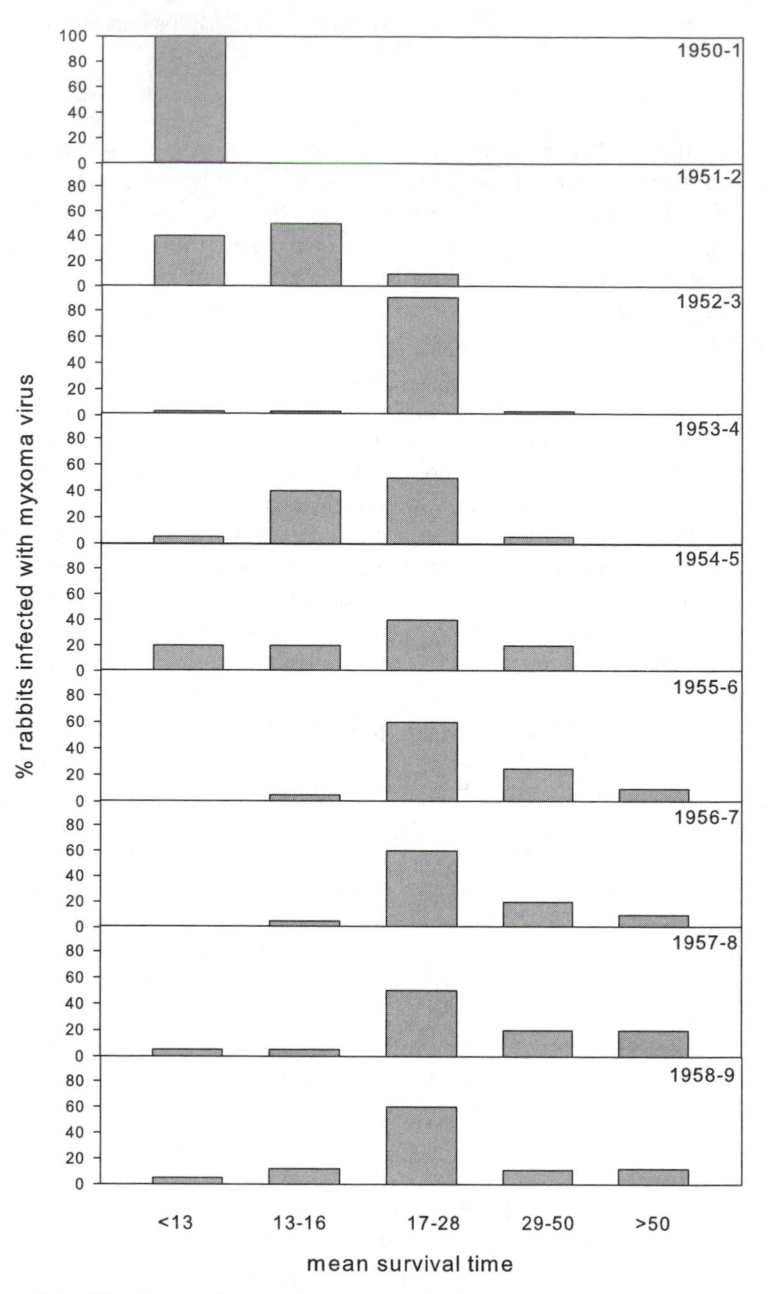

Figure 2-2 Virulence of myxoma virus toward rabbits of standard susceptibility, quantified by the mean number of days that the rabbits survived. The viral isolates were collected from field populations of rabbits in Australia at yearly intervals after the virus was introduced in 1950. [Reproduced from figure 1 of Fenner (1983) with permission from the Royal Society]

its transmission via biting insects, especially mosquitoes, which feed exclusively on living hosts. Throughout most of the range of the rabbits in Australia, the mosquitoes are not active in the winter months and therefore only those virus genotypes which keep the host alive through the winter are available for transmission in the spring. All the highly virulent genotypes were eliminated in the first winter. Virus genotypes of low virulence generate very low titers in the rabbit blood and, as a result, are transmitted inefficiently and are selected against. In this system, therefore, amelioration has led to an antagonistic relationship of intermediate virulence but not an association beneficial to the rabbit.

The key factor, additional to amelioration, required for the evolutionary transition from an antagonistic to mutualistic relationship is that the parasite possesses traits advantageous to the host. Such a trait may be a by-product of amelioration. To illustrate, let us consider another viral infection, bacteriophage in bacteria. Bacteriophage display two alternative strategies: the virulent lytic pathway, in which the virus produces multiple progeny that are released by lysis (and death) of the bacterial cell; and the nonvirulent lysogenic pathway, in which the viral DNA is integrated into the bacterial chromosome or transformed into a low-copy plasmid and then copied with each round of the bacterial chromosome replication and cell division. An integrated phage prevents subsequent infection by related phage particles, thereby ensuring its propagation to all the descendants of the infected bacterial cell. As a by-product, the bacterial host and its progeny acquire resistance to phage attack. However, the antagonism is not completely eliminated from the relationship between bacteria and their lysogenic phage. The infected bacterial cells grow and divide "under the Sword of Damocles" with the possibility that the lytic pathway is induced and the bacterial host is killed at any time.

Associations between fungi and grasses provide persuasive evidence for the evolutionary transition from an antagonistic relationship to a symbiosis. The symbiosis comprises systemic infections of the shoots of grasses by the asexual fungus *Neotyphodium* which synthesize and accumulate toxic alkaloids. Animals that feed on fungal-infected grasses ingest these alkaloids and become debilitated and may die. For example, ergot and indole diterpene-type alkaloids present in endophytes of tall fescue and perennial ryegrass cause neurological disorders known as "the staggers" in cattle and sheep and also confer resistance to various insect and nematode pests (Schardl 1996). This symbiotic protection is particularly valuable because grasses as a group have very limited intrinsic chemical defenses against herbivory. Most grasses persist by a combination of the physical defense of silica and a tolerance of herbivory, attributable mostly to the well-protected position of the shoot meristem at the base of the shoot.

The symbiotic fungi are related to pathogenic fungi of the genus *Epichloë*, which also synthesize alkaloids. The pathogens are localized primarily at the inflorescence of the plant host and reproduce sexually. They are deleterious to the grass because their reproductive structures enclose the grass inflorescence, preventing it from developing, and the infected tiller is reproductively sterile. Because the symbionts are generally asexual, the fungus protects the entire shoot against herbivores while permitting normal flowering and seed set. Hyphae of the asexual fungi commonly colonize the seed and are, thereby, transmitted vertically. In other words, sexual reproduction of fungus and grass is mutually exclusive.

The evolutionary relationship between the sexual and asexual endophytic fungi has been resolved. It is strictly from pathogen to symbiont. Most of the asexuals are hybrids between two *Epichloë* species, and they have arisen multiple times (Moon et al. 2004). The obligate vertical transmission of the asexual fungus selects for amelioration of the endophytic fungi, since costly or otherwise antagonistic traits that reduce seed set depress the fitness of the fungus. This relationship demonstrates how both a trait advantageous to the partner and the mode of transmission are important determinants of the evolutionary transition from parasitism to symbiosis. It is also of note that the origin of the symbiotic fungi by hybridization suggests that this transition can be evolutionarily rapid.

The selection pressure for the evolution of endophytic fungi overtly beneficial to plants may be particularly strong in agricultural settings. There is now excellent evidence that the concentration and diversity of alkaloids is far lower in the endophytic fungi of many grasses in natural vegetation than in agronomic grasses. Linked to this difference, grasses in natural vegetation generally derive little protection from their fungi against native herbivores (Faeth and Sullivan 2003). It appears that the evolution of consistently beneficial endophytic fungi in the agronomic grasses is a very special case shaped by human activities, including the high grazing pressure of domestic animals which selects for herbivore defenses in the grasses. The implication of this conclusion is that certain overtly beneficial interactions have evolved since the advent of human agriculture, i.e., over a period of a few thousand years or less.

The rapid transition from antagonism to symbiosis has been reported in a multiyear experiment on the bacterium *Wolbachia* in the fruit fly *Drosophila simulans* (Weeks et al. 2007). *Wolbachia* was first detected in Californian populations of *D. simulans* in 1985. Between 1985 and 1994, it was deleterious, causing 10–20% reduction in egg production by the flies. Despite the reduced fitness of *Wolbachia*-infected flies, *Wolbachia* spread through the population, driven by cytoplasmic incompatibility

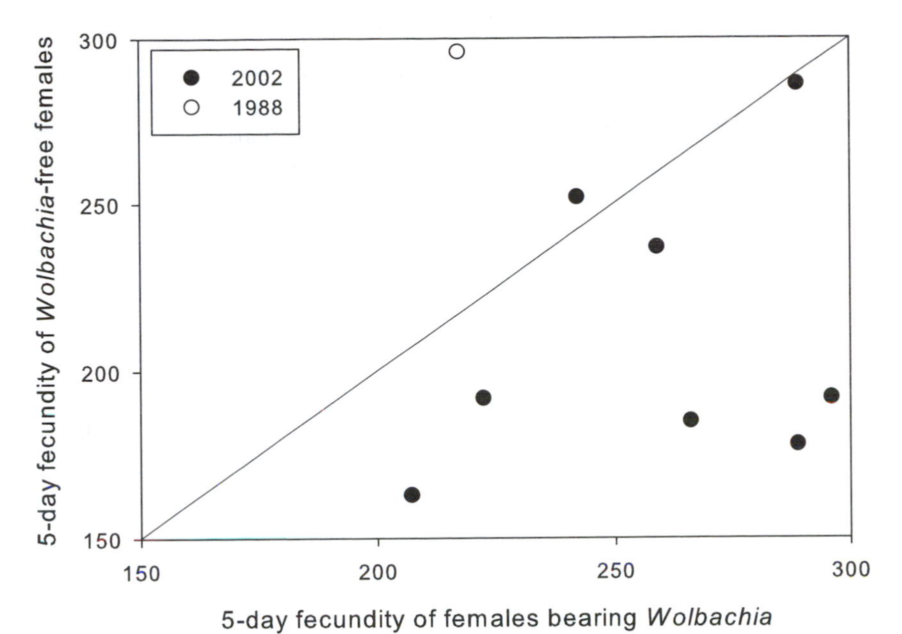

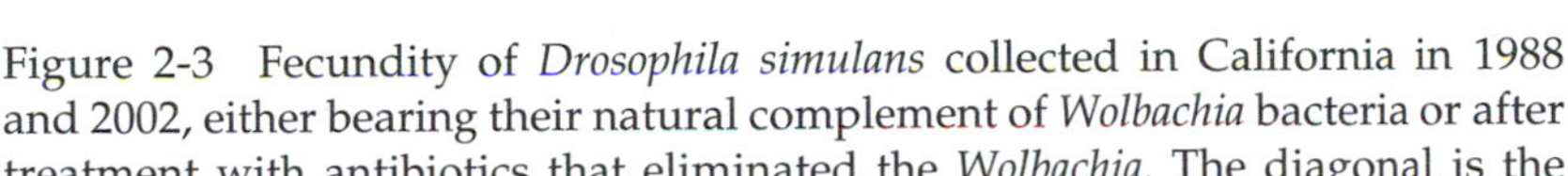

Figure 2-3 Fecundity of *Drosophila simulans* collected in California in 1988 and 2002, either bearing their natural complement of *Wolbachia* bacteria or after treatment with antibiotics that eliminated the *Wolbachia*. The diagonal is the line of equivalent performance of insects bearing and lacking *Wolbachia*. [Redrawn from figure 1A of Weeks et al. (2007)]

(CI). [In CI, matings between infected males and uninfected females are sterile, giving infected females a strong selective advantage. CI is described in more detail in chapter 3 (section 3.5.1).] Once its frequency in the host population is high, *Wolbachia* derives little advantage through CI and its fitness depends on the fecundity of the female host. As predicted, the fecundity of *Wolbachia*-infected females collected from the field in 2002 was either unaffected or reduced when *Wolbachia* was experimentally removed (figure 2-3). Supplementary experiments demonstrated that this change could be attributed to genetic changes in the *Wolbachia* and not the insect. It remains to be established how *Wolbachia* promotes host fitness. One possibility is that, as with the *Wolbachia* lineages in nematodes that I discuss below (section 2.2.6), the *Wolbachia* is nutritionally valuable to its *D. simulans* host.

2.2.4 *Addiction*

Addiction, like amelioration considered above, has the potential to contribute to the evolutionary transition from antagonism to symbiosis.

Addiction can be defined as dependence without benefit: an organism in an addictive symbiosis requires its partner for a level of fitness achieved by other organisms without the partner. It is potentially important in the evolutionary transition from parasitic to mutualistic associations. Because the organisms in an addictive relationship are dependent on each other, subsequent mutations that are beneficial for the partner are likely to be selected for, resulting in the transition to a mutually beneficial relationship.

The research of Jeon and colleagues on the amoeba *Amoeba proteus* provides an excellent illustration of addiction. Jeon's laboratory cultures of amoebae became infected with a bacterium that has never been identified and is known informally as the x-bacterium. The immediate effect of infection on the amoebae was negative, including reduced size and growth rates and increased mechanical fragility of the amoebae. However, within 200 generations, the amoebae became dependent on the bacterial infection, dying when the x-bacteria were eliminated with antibiotics. This transition from antagonism to dependence was experimentally repeatable, occurring reliably when naive amoebae (i.e., never previously exposed to x-bacteria) were infected with x-bacteria (Jeon and Ahn 1978). One difference between the naive and infected amoebae has been identified (Jeon and Jeon 2003); the gene coding for the enzyme S-adenosyl methionine synthetase (SAMS) was irreversibly repressed in infected amoebae within 50 days of infection, such that the amoebae were dependent on the SAMS enzyme activity in the bacteria for their supply of S-adenosyl methionine (which is a key methyl donor in metabolism). In other words, the x-bacteria rescue *A. proteus* from the negative impact of their own infection.

Another addictive symbiosis studied in detail concerns the bacterium *Wolbachia*. In insect hosts, *Wolbachia* is generally a reproductive parasite, as considered above (see section 2.2.3; also see chapter 3, section 3.5.1) but, exceptionally, it is required by the parasitic wasp *Asobara tabida*. The *Wolbachia* cells are transmitted vertically via the eggs but, when they are eliminated experimentally by antibiotic treatment, *A. tabida* is reproductively sterile with egg production halted at mid-oogenesis by massive apoptosis (programmed cell death) of nurse cells in the ovary (Pannebakker et al., 2007). A plausible evolutionary explanation provided by the authors for this effect is that normal oocyte development is accompanied by limited apoptosis, *Wolbachia* inhibits apoptosis, and *A. tabida* has responded to the *Wolbachia* infection by compensatory increase in the apoptotic signaling pathway. As a result, *A. tabida* is dependent on the inhibitory effect of *Wolbachia* for normal levels of apoptosis and the compensatory changes in *A. tabida* are harmful in the absence of *Wolbachia*.

Addiction may also be important in relation to the evolution of symbioses from nonantagonistic origins, as I consider below (section 2.4.1). Thus, although addiction may be one route for the evolutionary transition from antagonism to mutualism, not all addictive symbioses are necessarily at a transitional stage between antagonism and mutualism. A further issue is that very few instances of addictive symbioses have been described in the literature, but it is unclear whether this is because they are rare or unrecognized. Further research is required to assess the general significance of addiction in the evolutionary transition from antagonism to mutualism.

2.2.5 From Mutualism to Antagonism

Much discussion of the evolutionary stability of mutualisms has focused on the selection pressure on each partner to exploit the association to maximize its individual fitness (e.g., Herre et al. 1999; Sachs and Simms 2006). As discussed in chapter 1 (section 1.4.1), some strands in the literature have concluded that mutualisms are inherently unstable, with a tendency to evolve into parasitic relationships.

There is excellent evidence for organisms within symbioses that fail to provide services and I discuss these in chapter 3 (section 3.3). Despite this, there are few unambiguous examples of parasitic species with symbiotic ancestors apart from the various bacteria which gain and lose virulence via horizontally mobile pathogenicity islands (see section 2.2.1). Most of the symbiosis-derived parasites are obligate parasites of the ancestral symbiosis, i.e., they depend on the persistence of the symbiosis that they parasitize. For example, some plants parasitize mycorrhizal fungi that are also linked to other plants. In this way, the parasites gain access to inorganic nutrients, such as phosphate, from the fungus and carbon fixed by other plants tapped into the same mycorrhizal network. An estimated 400 species of acholorophyllous plant parasites from 10 families obtain their total carbon requirement from mycorrhizas (Taylor and Bruns 1997; Bidartondo et al. 2002). Other plants, including various woodland orchids living in deep shade, meet part of their carbon requirements by exploiting mycorrhizas (Bidartondo et al. 2004). These parasites are called mycoheterotrophs, defined as plants which derive fixed carbon from fungi. (Not all mycoheterotrophic plants are mycorrhizal parasites; some associate with saprotrophic fungi.) Other parasites of symbioses include nonpollinating insect species derived from seed predator pollinators, including fig wasps and yucca moths, which are discussed in chapter 3 (section 3.3).

Parasites that are independent of related mutualistic species have evolved among the lycaenid butterflies. The ancestral condition of the family Lycaenidae was almost certainly in association with tending ants. The lycaenid caterpillar feeds on plants. It secretes sugar-rich solution from a specialized gland, the Newcomers's gland on its dorsal surface; ants feed on this secretion and protect the caterpillar from predators. An estimated 200 (4%) of the ~5000 lycaenid species parasitize their ant partners, and the parasites have evolved independently several times. The best-studied parasites are the *Maculinea-Phengaris* clade, which exploits myrmecine ants (Als et al. 2004). The early instar caterpillar of the parasitic lycaenids has a mutualistic relationship with the ants but, on reaching the fourth instar, it drops from the plant to the ground, where it is picked up by the ants. The caterpillar bears surface hydrocarbons that mimic the surface chemistry of ant larvae (Nash et al. 2008), prompting the ants to transport the caterpillar to their nest. Many *Maculinea* species feed voraciously on the ant larvae (for example, a single caterpillar of the large blue butterfly *M. arion* can decimate its host ant colony), but other species, e.g., *M. rebeli*, are "cuckoos," feeding on regurgitant from ant workers. The *Maculinea* species pupate in the ant nest over the winter and emerge as adults in the early summer. As the full lifecycle reveals, transition to antagonism in *Maculinea* species is not complete, because the ancestral mutualistic lifestyle is retained in the young larvae.

Why is mutualism-derived parasitism apparently so rare? There are two explanations. The first is that mutually beneficial relationships are robust to between-partner conflict, with effective mechanisms for conflict resolution. The routes by which conflicts are resolved are addressed in chapter 3 (section 3.4). The second possible explanation is that antagonistic relationships evolve frequently from mutualisms but they are evolutionarily unstable. Many associations might go extinct, or the exploited partner might either switch to a different mutualist or revert to the free-living condition. Consistent with this possibility, Pierce et al. (2002) argue that the parasitic lycaenid clades are prone to extinction. Also, some of the ~25% of all lycaenids that are secondarily independent of ants might have been abandoned by their ant partners in response to parasitic tendencies of the lycaenid. Systematic analysis of the incidence of extinction in groups which include mutualistic taxa, including the lycaenid clades with different lifestyles, would help to resolve these issues.

2.2.6 Evidence from the Phylogenetic Relationships Between Symbionts and Parasites

For most of the associations explored in the preceding sections, we can have some confidence in the direction of the evolutionary transition (from antagonism to mutualism or vice versa). Various other groups include closely related parasites and symbionts, but it can be difficult to interpret the pattern of lifestyle evolution from phylogenetic data. This is illustrated by studies of some bacteria.

There are two functional groups of bacteria in the genus *Wolbachia*: obligate symbionts of filarial nematodes and reproductive parasites of arthropods, especially insects. It is plausible that *Wolbachia* have been transferred between the two groups of hosts because various insects act as vectors for the transmission of filarial nematodes between mammalian hosts. However, the nematode and arthropod-affiliated *Wolbachia* are phylogenetically distinct, pointing to a single or few evolutionary transfers between the two groups of hosts. In which direction? The deeper branches for the phylogeny of nematode-associated *Wolbachia* have led Fenn and Blaxter (2004) to favor *Wolbachia* transfer from nematode to arthropod. Transfer in the reverse direction cannot be excluded, however, since the branch lengths could be distorted by higher rates of sequence evolution in the nematode-affiliated *Wolbachia*. Also, the evidence for insects both addicted to and benefiting from *Wolbachia* (see sections 2.2.3 and 2.2.4) indicates the feasibility of an evolutionary transition from parasite to symbiosis. An additional caveat in interpreting the phylogenetic data is that the lifestyle of the common ancestor and stem lineages of *Wolbachia* are unknown, raising the possibility that the obligate symbionts and reproductive parasites may have evolved independently from a common ancestor with a predisposition to associate with animals.

The Chlamydia is an ancient group of obligate intracellular bacteria that was traditionally considered as exclusively pathogenic. Nevertheless, phylogenetic analyses have revealed that Chlamydia occur in amoebae and insects, where they are harmless or beneficial (Horn et al. 2004). The animal parasites are most unlikely to be representative of the ancient lifestyle. Molmeret et al. (2005) propose that the common ancestor of all modern Chlamydia were nonpathogenic inhabitants of amoebae or other protists, in which interactions with eukaryotes evolved; and that these amoeba-associated Chlamydia were the immediate progenitors of both the pathogens and symbionts of animals. The evolutionary transitions may have been more complex than this, potentially involving multiple transitions between pathogenic, harmless, and symbiotic lifestyles.

Let us take stock. I have described close relationships between parasites and symbionts in various phylogenies, including the Chlamydia and *Wolbachia* (above), *Cordiceps* (section 2.2.2), and *Epichloë*/*Neotyphodium* (section 2.2.3). These data collectively indicate that taxa with predispositions for intimate interactions with living eukaryotic cells can adopt various different lifestyles. In many cases, the direction of evolutionary transitions is unclear and, for some groups, it is probably variable. This diversity is illustrated particularly well by the nitrogen-fixing rhizobia in root nodules of leguminous plants. (The term rhizobia is taken to mean any bacteria that form nitrogen-fixing symbioses in legume nodules.) The rhizobia have multiple origins, mostly within the α-proteobacteria but also including members of β-proteobacteria and perhaps γ-proteobacteria (e.g., Moulin et al. 2001). The sister taxa of these different rhizobia include symbionts and free-living soil bacteria, as well as pathogens. As already discussed (section 2.2.1), *Rhizobium* and related genera (e.g., *Mesorhizobium*, *Ensifer*) are related to the plant pathogen *Agrobacterium* and animal parasites including *Brucella*; while the second major group of α-proteobacteria, comprising *Bradyrhizobium/Azorhizobium*, has no known pathogenic relatives and is allied with free-living bacteria including the photosynthetic *Rhodopseudomonas* (Lafay and Burdon 1998). The β-proteobacterial rhizobia include members of the genus *Burkholderia*, which also has symbiotic representatives (e.g., within the hyphae of arbuscular mycorrhizal fungi and in insects), opportunistic pathogens, and free-living taxa (Vandamme et al. 2002). The implication of these data on the rhizobia is that bacteria which interact mutualistically with their hosts and perform broadly uniform symbiotic functions can have very different evolutionary histories.

Equally striking, however, are the many groups of parasites without known mutualistic relatives. These include various animal parasites (e.g., tapeworms, trematodes, ticks), plants (e.g., *Striga* and *Orobranche*), and protists (e.g., microsporidia, diplomonads). We must therefore look to other sources for the origin of some symbiotic organisms. One important source in this symbiosis-rich world is preexisting symbioses, as is considered in the next section.

2.3 Origins of New Symbioses by Partner Capture

For at least three major groups of symbioses, there is persuasive evidence that at least one of the progenitors of an evolutionarily novel symbiosis has an evolutionary history of symbiosis. The lineage of symbiotic organisms has, in this way, captured a novel partner. These three examples of partner capture relate to dinoflagellate algae, mycorrhizal fungi, and microbial symbionts of insects, and I consider them here.

2.3.1 Partner Capture Among Symbiotic Algae

Symbiotic algae have undoubtedly acquired novel hosts. One example is the dinoflagellate *Symbiodinium*, the usual symbiont of benthic cnidarians (corals, sea anemones, etc.) in the photic zone at low latitudes.

Symbiodinium is also found in tridacnid clams, clionid sponges, and certain acoel flatworms. The phylogenetic distribution of *Symbiodinium* is most readily explained by the evolutionary expansion of the host range of *Symbiodinium*, most probably from corals and related cnidarians to other animals. It is not known whether the novel hosts colonized by *Symbiodinium* were alga-free or bore other algae which were replaced by *Symbiodinium*; the usual photosynthetic symbionts in nonclionid sponges are cyanobacteria, and acoel flatworms with prasinophyte algae and diatoms are known. The current distribution of freshwater *Chlorella* symbionts in various protists, sponges, hydras, flatworms, and freshwater clams is also explained most parsimoniously by the evolutionary capture of novel hosts by symbiotic *Chlorella*.

Phylogenetic data suggest that the ancestral symbiosis in scleractinian corals might not have involved *Symbiodinium*. There is fossil evidence that scleractinian corals had acquired photosynthetic symbionts ~200 million years ago in the late Triassic (see chapter 5, section 5.2.1) but molecular evidence that the common ancestor of modern *Symbiodinium* arose either 65 million years ago at the K/T boundary (Tchernov et al. 2004) or more recently ~50 million years ago in the early Eocene (Pochon et al. 2006). One possible explanation for this discrepancy is that an ancestral symbiont of corals was replaced by *Symbiodinium* at 65/50 million years ago. Alternatively, the Mesozoic corals may have borne dinoflagellate symbionts which underwent a severe bottleneck at 65/50 million years, followed by rapid diversification of *Symbiodinium* in modern corals.

Dinoflagellates are hosts as well as symbionts. As hosts, they appear to be predisposed to switch partners. Many dinoflagellates, including *Symbiodinium*, have a characteristic peridinin-containing plastid bounded by three membranes, but this plastid has been replaced multiple times by various algae, including haptophytes, cryptophytes, prasinophytes, and diatoms (Saldarriaga et al. 2001).

2.3.2 Partner Capture in Mycorrhizas

Partner capture and exchange have been crucially important in the evolutionary history of mycorrhizas. Several types of mycorrhizas are recognized (figure 2-4). Of these, the arbuscular mycorrhizas (AMs) are the most ancient, probably evolving in early land plants about 400 million years ago in the Devonian and widely distributed today among the bryophytes (especially liverworts and hornworts), pteridophytes, gymnosperms, and angiosperms. There is strong evidence that plants have subsequently captured alternative fungal partners, often accompanied by the elimination of AM fungi, giving rise to functionally and morphologically

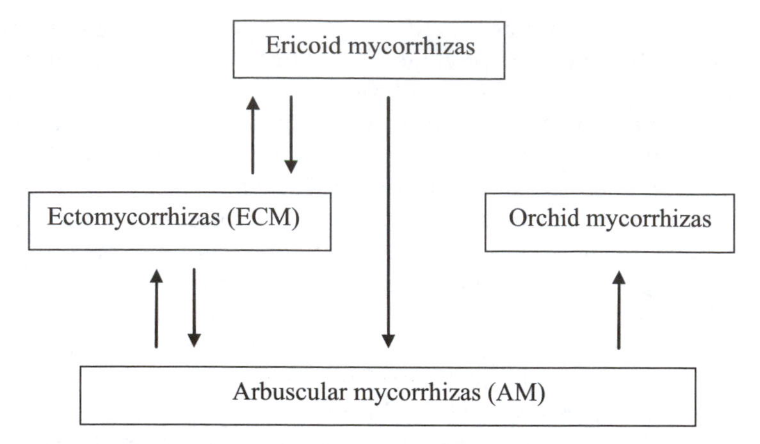

Figure 2-4 Evolutionary transitions of mycorrhizas among plants (see text for details). Note that all orchids are associated with orchid mycorrhizal fungi during seedling growth but many species switch to AM fungi or ECM fungi at later developmental stages.

distinct mycorrhizas. I will consider the displacement of AM fungi from plant roots first, and then the evolutionary origin of the AM.

The ectomycorrhizal (ECM) fungi are predominantly basidiomycetes related to saprotrophic fungi, especially wood-decaying taxa. They probably arose at the time of the evolutionary origin of the gymnosperm Pinaceae (ca. 120 million years ago) and the angiosperm Fagales (ca. 100 million years ago) in the Cretaceous (Brundrett 2002). The transition from AM to ECM may have been a slow process, involving plants colonized by both AM fungi and ECM fungi as evolutionary intermediates, and it is not irreversible. This is indicated by the various plants today (e.g., *Alnus*, *Salix*) which can be infected by both types of mycorrhizal fungi and the phylogenetic evidence that some plant lineages, especially in the Myrtales and Fabales, have switched back and forth multiple times between ECM and AM (figure 2-4). Ecological factors may contribute to the predisposition of certain plant groups to associate preferentially with ECM fungi. In particular, the dominant ECM plants include many woody perennials growing in seasonal climates at high latitudes on soils containing organic nitrogen. ECM fungi are generally better suited than AM fungi to exploit these conditions because they form a substantive sheath around roots in which seasonally available nutrients can be stored (Smith and Read 2007).

Two distinctive groups of mycorrhizas are taxonomically restricted: the ericoid mycorrhizas in the Ericales (Ericaceae and Epacridaceae), and the orchid mycorrhizas in the Orchidaceae. The evolutionary history of mycorrhizas in the Ericales is complex. The ancestral condition was probably AM—at least, the basal genus *Clethra* is AM (Brundrett 2002)—but the AM fungi have been replaced widely by ECM fungi. Phylogenetic data point to a single evolutionary origin of the ericoid mycorrhizas involving the replacement of ECM fungi by saprotrophic ascomycetes, probably in the late Cretaceous ~80 million years ago (Cullings 1996; McLean et al. 1999). This transition may have been linked to ecological factors, with the ericoid mycorrhizal fungi able to utilize acidic soils rich in toxic metals, as found in heathlands today (Smith and Read 2007). These evolutionary transitions are reversible, with evidence that some Ericales have reverted from the ericoid mycorrhizas to both ECM and AM (Koske et al. 1990) (figure 2-4).

From one perspective, the evolution of orchid mycorrhizas is simple: replacement of AM fungi by orchid mycorrhizal fungi, coincident with the evolutionary origin of orchids about 100 million years ago. Functionally, however, this replacement event must have been complex because the direction of carbon transfer in orchid mycorrhizas is from fungus to plant, the reverse of that in AM. The fungal-derived organic carbon is central to the evolutionary origin and diversification of orchids. A key feature of orchids is their ability to exploit specialized and very patchy habitats, made possible by their production of dust seeds, i.e., very tiny (one to several micrograms) seeds devoid of any reserves to support growth of the germinating seedling. This strategy is linked to the receipt of organic carbon from the mycorrhizal fungi. All orchids, as a consequence, are dependent on their mycorrhizal fungi for germination and seedling growth under natural conditions. Some orchid species are achlorophyllous, remaining dependent on the mycorrhiza for organic carbon throughout their lives, but many develop into photosynthetic plants, and become independent of their orchid mycorrhizal fungi. (Some photosynthetic orchids associate with AM or ECM fungi.) The evolutionary origins of the orchid mycorrhizal association are uncertain. Brundrett (2002) has argued that the ancestral relationship with AM fungi may have been partly exploitative, meaning that the AM fungal mycelium was also connected to other plants, and the mycorrhizal fungus was a conduit for the net transfer of organic carbon from those other plants to the ancestor of orchids. This general scenario is plausible because plants that acquire organic carbon from other plants through mycorrhizal fungi have evolved various times (see chapter 3, section 3.3). The ancestral orchid, however, appears to have been unique in the putative transition to a relationship with mycorrhizal fungi independent of other plants.

Let us turn now to the ancestral mycorrhizal relationship, the AM. The fungi are exclusively zygomycetes of the order Glomales, a phylogenetically isolated group with the common feature of short-lived dichotomously branching haustoria called arbuscules. The evolutionary origins of this symbiosis are obscure, and it has been suggested that the fungi evolved either from pathogens of early land plants (or their aquatic ancestors) or from persistent fungi in the plant tissues that consumed excess photosynthesis-derived carbon without causing undue harm to the plant. Data on a highly divergent glomalean genus, *Geosiphon*, suggest a different scenario. *Geosiphon* does not associate with plants, but with cyanobacteria, and the interface between the fungal hypha and cyanobacterial cell is structurally similar to the arbuscule interface in AMs (Schüssler and Kluge 2000). This raises the possibility that the capacity of glomalean fungi to form intimate symbioses with other organisms and to generate arbuscules predated the relationship with plants, and that the origin of the AM fungus–plant association can best be described as the capture of a new partner by a symbiotic fungus. If the AMs evolved by this route, then the arbuscule of the AM fungi is not an adaptation to the mycorrhizal relationship but carried over from a previous relationship with cyanobacteria or other photosynthetic partners.

2.3.3 Partner Capture Among Intracellular Microbial Symbioses of Insects

A further example of partner capture relates to a remarkable group of symbioses in insects and a few other arthropods (notably the ticks), known generically as mycetocyte symbioses. These relationships have evolved independently multiple times between different arthropods (including representatives of seven insect orders) and various bacteria and fungi (table 2-1). The mycetocyte symbioses share three traits.

- The microorganisms are intracellular, restricted to cells known as mycetocytes, whose sole function appears to be to house and maintain the microorganisms. (Some authors describe the mycetocytes that bear bacteria as bacteriocytes.)
- The microorganisms are obligately vertically transmitted from mother to offspring, usually via the cytoplasm of the egg.
- The association is required by both the arthropods and microorganisms.

The hosts of most mycetocyte symbionts live on nutritionally poor or unbalanced diets, suggesting that the microorganisms provide their hosts with nutrients, such as amino acids and vitamins (Buchner 1965), an interpretation that has been supported for several systems studied experimentally (Douglas 1998; Dale and Moran 2006).

TABLE 2-1
Mycetocyte Symbioses in Insects

Insect	Microorganisms
	(a) Plant sap feeders
Hemiptera	
Auchenorrhyncha (e.g., leaf-hoppers, plant-hoppers)	*Baumannia cicadellinicola* (γ-proteobacteria) and *Sulcia muelleri* (Bacteroidetes); Pyrenomycete fungi in some plant-hoppers
Aphids	*Buchnera aphidicola* (γ-proteobacteria) or fungi
Whitefly	*Portiera aleyrodidarum* (γ-proteobacteria)
Psyllids	*Carsonella ruddii* (γ-proteobacteria)
Scale insects and mealybugs	*Tremblaya princes* (β-proteobacteria)
	(b) Vertebrate blood
Hemiptera	
Cimicids (bedbugs)	Not known
Triatomine bugs	Not known
Anoplura (sucking lice)	*Riesia pediculicola* (γ-proteobacteria) in human head louse and body louse
Diptera Pupiparia	*Wigglesworthia* spp. in tsetse flies
	(c) Generalist feeders
Blattoidea (cockroaches)	*Blattabacterium cuenoti* (flavobacteria)
Mallophaga (biting lice)	not known
Psocoptera (book lice)	*Rickettsia* spp.
Coleoptera, including	
Curculionidae (weevils)	Various γ-proteobacteria
Anobiidae (timber beetles)	*Symbiotaphrina* (yeasts)
Hymenoptera	
Camponoti (carpenter ants)	*Blochmannia* spp. (γ-proteobacteria)

Mycetocyte symbioses are very widespread or universal in four insect orders: the Blattoidea (cockroaches), Hemiptera (sucking bugs), Anoplura (sucking lice), and Coleoptera (beetles, for which they are either widespread or universal in ten different families); they also occur, although with a very restricted distribution, among the ants in the Hymenoptera and in the Diptera (including tsetse flies *Glossina*) and are present in a single termite species *Mastotermes darwiniensis*. [Details, including references are provided in Douglas (2007)]

The mycetocyte symbionts have probably evolved from members of the gut microbiota or from vertically transmitted microorganisms with a broad tissue distribution (Douglas 1989), and most are not related to pathogens (Moran and Wernegreen 2000). Furthermore, there is good phylogenetic evidence that some insects with an evolutionary history of mycetocyte symbioses have captured novel microbial partners. For example, three different groups of γ-proteobacteria have been identified in dryophthorid weevils. From the distribution of these bacteria among ten dryophthorid genera tested, Lefèvre et al. (2004) concluded that R-clade is basal and has been replaced on two separate occasions, once by S-clade and once by D-clade. These transitions can be linked with diet: those with R-clade feed on monocots, those with S-clade are grain-feeders, and those with D-clade consume decaying wood. This raises the possibility that the different bacteria make different nutritional contributions that are suited to different dietary regimes.

I have already introduced the usurping microorganism in aphids. It is the *Cordiceps* fungus that is related to the symbionts of some planthoppers and, more distantly, various pathogens (section 2.2.2). *Cordiceps* occurs in just one aphid subfamily, the Cerataphidinae; other aphids bear the γ-proteobacterium *Buchnera*. The switch from *Buchnera* to *Cordiceps* is not known to be linked to dietary differences between the Cerataphinidae and other aphids. However, *Buchnera* displays some features predicted of genomic deterioration, and the collapse of its effectiveness through genomic meltdown in certain lineages could have precipitated its displacement by other bacteria. (The causes of genomic deterioration in microbial symbionts subject to persistent vertical transmission are considered in chapter 3, section 3.5.3.)

An experimental analog of the evolutionary replacement of symbionts is provided by the research of Koga et al. (2003) on pea aphids. Some pea aphids bear a bacterium *Serratia symbiotica* in addition to *Buchnera*. *S. symbiotica* is not required by the aphid and tends to depress aphid growth and reproduction (figure 2-5a). Koga and colleagues treated the aphids with antibiotics to eliminate all the bacteria, including *Buchnera*, and these aphids produced no offspring. However, when these bacteria-free aphids were reinfected with *S. symbiotica* alone, they produced a few offspring and persisted over multiple generations (figure 2-5b). In a similar fashion, the ancestor of the *Cordiceps* symbiont in cerataphinid aphids and S-clade/D-clade symbionts in weevils could have rescued their hosts from a poorly functioning *Buchnera* and R-clade symbiont, respectively, and subsequently evolved into effective symbionts under the selection pressure of vertical transmission. As with *S. symbiotica*, the usurping *Cordiceps* and S/D-clade bacteria may have been resident in the ancestral insect host at low densities.

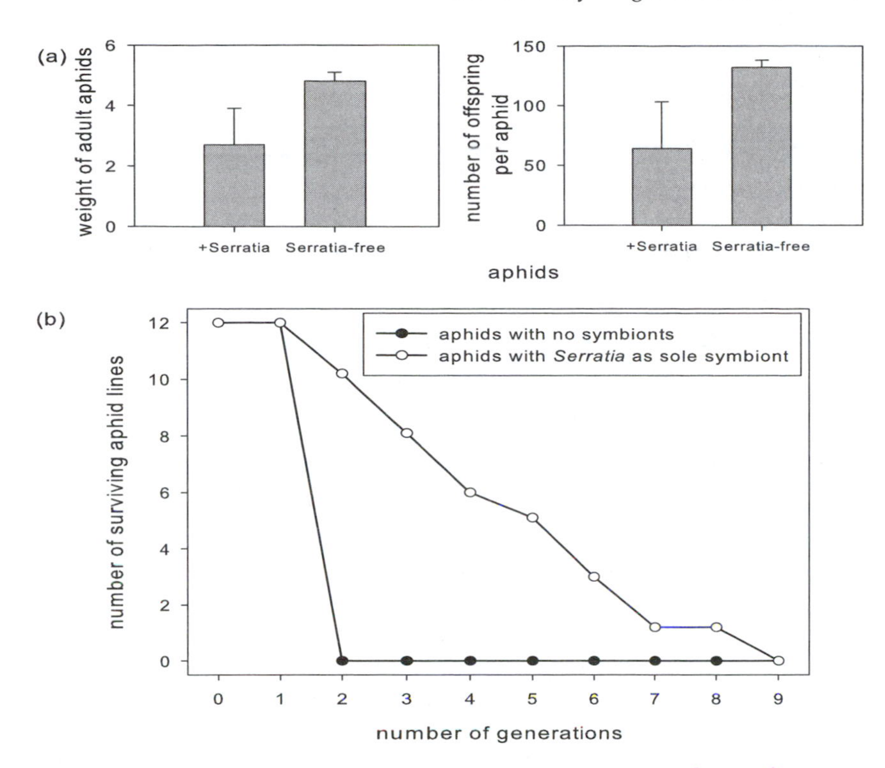

Figure 2-5 Impact of the bacterium *Serratia symbiotica* on the performance of the pea aphid *Acyrthosiphon pisum*. (a) Weight and reproductive output of aphids bearing the bacterial symbiont *Buchnera aphidicola*. (b) Persistence through nine asexual generations of aphids from which *B. aphidicola* has been eliminated. [Redrawn from figures 1 and 6b of Koga et al. (2003)]

2.3.4 *Partner Capture and Shifts in Allegiance*

Many mutualisms involve an organism that protects its partner against an antagonist. In certain instances, the protector can change allegiance to become the partner of the antagonist. Formally, this type of evolutionary change can be described as both an evolutionary shift between mutualism and antagonism (in both directions simultaneously), which was considered in section 2.2, and as capture of a novel mutualistic partner addressed here.

Changing allegiance has been documented particularly clearly for the ant *Lasius niger*. This ant tends various aphids, protecting them from predators and parasitic wasps. Contrary to this generality, colonies of the black bean aphids *Aphis fabae* feeding on thistle *Cirsium arvense* suffer a higher incidence of parasitization by the wasp *Lysiphlebus*

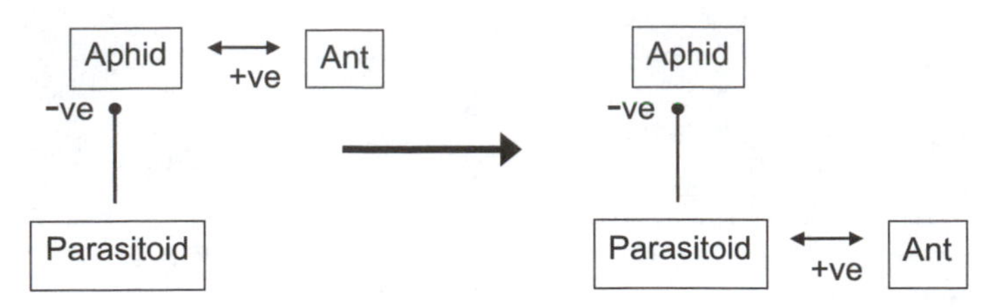

Figure 2-6 Partner switching by the ant *Lasius niger* involves the evolutionary transition of the ant from a mutualism with the aphid *Aphis fabae* to a mutualism with the parasitoid *Lysiphlebus carduii*.

cardui when they are tended by *L. niger* than when isolated from tending ants (Völkl 1992). The female *L. cardui* wasps inject eggs into the aphids which remain alive and feeding for some days, supporting the growth and development of the wasp eggs and larvae. The failure of the ants to protect the aphid colonies from *L. cardui* is to the ants' advantage because recently parasitized aphids produce more honeydew than unparasitized aphids. In effect, the ants are in a mutually beneficial interaction with the parasitoid: the ants obtain more food, while the parasitoids gain open access to aphids that continue to be protected by the ants from supernumerary attack by other predators or hyperparasitoids. The ants have shifted allegiance from a mutualism with the aphids to a mutualism with the parasitoid (figure 2-6).

2.4 Making and Breaking Symbioses

2.4.1 From Chance Encounter to Symbiosis

Many organisms in symbioses have evolved from a different type of relationship, either antagonistic (section 2.2) or a symbiosis with a different partner (section 2.3). Here, I address a further evolutionary origin, directly from the nonsymbiotic state. Organisms with no history of interaction can come together to their mutual benefit if each possesses a trait that, under particular circumstances, is advantageous to the other organism.

The feasibility of symbiosis through chance encounter is illustrated by the experimental study of Christensen et al. (2002) on two bacteria, an *Acinetobacter* sp. and *Pseudomonas putida*. Both of these bacteria can utilize benzoyl alcohol as a sole carbon and energy source, but *Acinetobacter* sp. can utilize it more efficiently than *P. putida*, and *P. putida* can metabolize benzoate, an intermediate in the degradation of benzoyl alcohol, more efficiently than *Acinetobacter*. The relationship between the two bacteria depends critically on the culture conditions, specifically whether the cells are grown in suspension, precluding close contact, or grown on a surface, where they can form colonies. When maintained in suspension, *P. putida* and *Acinetobacter* compete for the benzoyl alcohol; their relationship is antagonistic, resulting in the suppression of *P. putida* (figure 2-7a). When grown on a surface, the two bacteria form stable mixed biofilms, each comprising a microcolony of *Acinetobacter* sp. on the surface of which *P. putida* cells accumulate. In these assemblages, the benzoyl alcohol is degraded principally by *Acinetobacter* sp. and the intermediate degradation product, benzoate, diffuses rapidly to the nearby *P. putida* cells, which complete the degradation. The bacteria growing as biofilms on a surface consume benzoyl alcohol more rapidly and grow faster in mixed culture than in isolation (figure 2-7b), indicating that they are interacting to their mutual advantage. The opportunity for cross-feeding of metabolites between the two bacterial taxa is created by the complementarity of their preexisting traits and the particular circumstances of close proximity, i.e., as a symbiosis.

Chance encounters between microorganisms and both animals and plants are commonplace because of the ubiquity of microorganisms. Many of the interactions are transient and of no significance to either participant. Such casual relationships can be invoked as the evolutionary progenitors of symbioses, especially where the organisms in symbiosis lack parasitic ancestors or relatives. For example, detailed molecular phylogenies of basidiomycete fungi reveal that the ectomycorrhizal fungi associated with plants have arisen from saprotrophic soil fungi (Bruns and Shefferson 2004), and origins from casual contact between fungal hyphae and plant roots are plausible. Other microbial symbionts also lack parasitic relatives, including the chemoautotrophic bacteria in marine invertebrates (e.g., pogonophoran worms and bivalves associated with deep-sea hydrothermal vents), some intracellular bacterial symbionts, such as *Buchnera* and *Wigglesworthia* in insects, and symbiotic algae, e.g., *Chlorella* and *Symbiodinium* in animals and *Trebouxia* in lichenized fungi.

Chance encounter may also have played a strong part in the evolution of some interactions between animals and plants, including interactions with partner contact too brief to be considered as symbioses. One instance is the evolution of pollination by the neotropical euglossine bees. Unlike most pollination systems, this interaction is not based on food collection by the animal, but the collection of perfumes. An estimated 650 orchid species are absolutely dependent on euglossine bees for pollination. These plants attract male bees by providing organic volatiles, which are collected by the male bees into a perfume pocket on

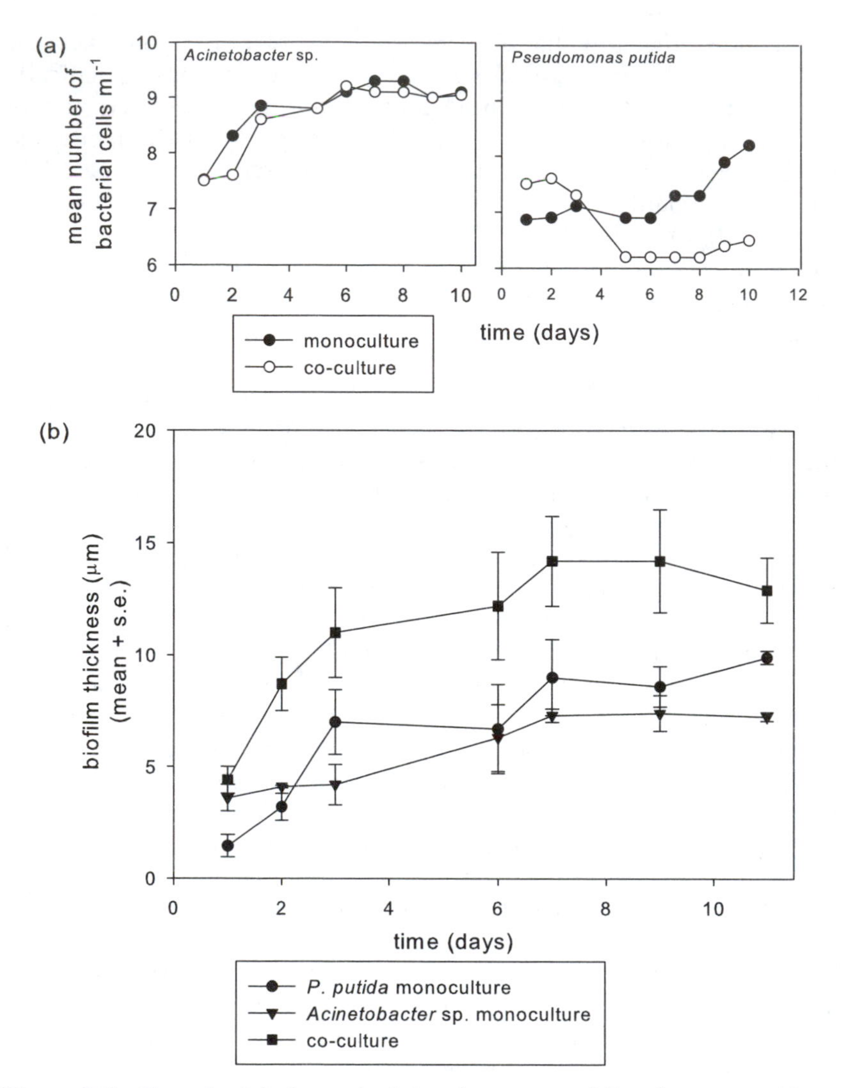

Figure 2-7 Growth of the bacteria *Acinetobacter* sp. and *Pseudomonas putida* on benzoyl alcohol as sole carbon source. (a) In suspension, the bacteria in co-culture competed and the growth of *P. putida* was suppressed. (b) On a surface, the bacteria formed biofilms and growth was promoted by co-culture. [Reproduced from figures 1 and 2 of Christiansen et al. (2002) with permission from the American Society for Microbiology]

the hind tibia, concentrated by extraction with a lipophilic substance, and then released from the pocket to attract females for mating (Eltz et al. 2007). Scent is used widely by other insects, including some bees, for mate attraction, and the relationship between euglossines and orchids is likely to have evolved by a switch from intrinsic scents (synthesized by the bee) to orchid-derived scents, presumably through chance encounter with the plants.

Addiction may have contributed to the evolution of persistent symbioses between organisms that have made contact by chance. I have introduced the concept of addiction (i.e., dependence without benefit) earlier in this chapter in the context of the evolutionary transition of antagonistic to mutualistic associations (section 2.2.4). Addiction can also evolve in an organism which is assured of contacting a partner that is selectively neutral (i.e., is neither beneficial nor deleterious to the organism). In some respects, mammals are addicted to their gut microbiota. The normal development of the capillary network in the small intestine is dependent on the presence of gut bacteria. When mice are reared under aseptic conditions so that the digestive tract is sterile, the small intestine is very poorly vascularized and the mice have a much reduced capacity to assimilate nutrients from the gut. Capillary growth is not induced by the bacteria directly, but by antibacterial peptides which are produced by specific intestinal cells, known as Paneth cells, in response to the presence of these bacteria (Stappenbeck et al. 2002). In this way, the bacteria are a necessary component in an animal signaling pathway essential for normal development of the capillaries, but the intestinal capillary beds function independently of the bacteria. There is, presumably, no selection pressure on mammals to eliminate this addiction because they are invariably infected by gut microorganisms at birth or soon afterward. The gut microbiota is also beneficial to their mammalian hosts by the provision of nutrients, such as vitamins, and protection from pathogens (Wilson 2005). Addiction may often be evolutionarily ancient, providing the sustained contact required for the subsequent evolution of beneficial interactions; and it may also promote the persistence of symbioses under conditions where the value of beneficial interactions is small.

2.4.2 The Evolutionary Breakdown of Symbioses

Our understanding of the breakdown of symbioses is very limited, in terms of its incidence, the predisposing factors, and evolutionary consequences for the organisms involved. Both between-partner conflict and external factors (e.g., change in environmental conditions, nutrient availability) can be invoked as explanations for symbiosis breakdown, but the relative importance of these factors and their interactions in precipitating the collapse of associations is uncertain. Similarly, the fate of participants in symbioses that collapse is uncertain; they could go extinct, adopt a free-living lifestyle, or switch to a different symbiotic partner. These issues can only be resolved fully by systematic analysis, but some relevant data are already available on the evolutionary transition of symbiotic organisms to the free-living condition.

Instances of symbiosis collapse have been known for decades and are not controversial. For example, several families of dicot angiosperm plants (Brassicaceae, Carypophyllaceae, Chenopodiaceae, and Uricaceae) have lost the capacity to form symbioses with mycorrhizal fungi (Brundrett 2002). Other putative examples have emerged from detailed multitaxon phylogenies that have placed nonsymbiotic taxa embedded within clades of symbiotic organisms. This is particularly evident for partners in lichen symbioses, with several major clades of free-living ascomycete fungi (and their asexual derivatives) nested within lichenized fungal clades (Lutzoni et al. 2001) and free-living taxa of cyanobacteria derived from lichen symbionts (O'Brien et al. 2005). These proposed transitions are entirely congruent with the evidence that some lichenized fungi can persist as microcolonies in isolation (Wedin et al. 2004). There are, however, difficulties with the design and interpretation of some large, single-gene phylogenies and the conclusions about the evolutionary relationships of some symbiotic organisms are misleading (Bruns and Shefferson 2004). As considered in the opening paragraphs of this chapter (section 2.1), improved algorithms and multigene (including whole genome) analyses may lead to revisions of some phylogenies, with implications for our understanding of the incidence and evolutionary consequences of symbiosis breakdown.

Ant associations with animals and plants show some propensity for evolutionary breakdown. Lycaenid butterflies have lost the association with ants multiple times (see section 2.2.5). Ant associations are also exceptionally labile in hemipteran insects, with multiple examples of single genera comprising species that are obligately tended, facultatively tended, and not tended by ants. A reconstruction of the phylogeny of just 15 of the 88 species of the aphid genus *Chaitophorus* identified multiple losses (and gains) of association with ants and, furthermore, linked ant-tending to feeding habit. Specifically, species with long mouthparts that feed from deep sieve elements in their host plants are vulnerable to predators because of the extended time required to withdraw their stylets, and they are particularly predisposed to form and retain associations with ants; while species with more shallow feeding sites derive relatively small or variable benefits of ant-tending and are prone to lose the association. In these associations where shifts in ant-tending status can follow evolutionary shifts in feeding traits of aphids, selection pressures on the plant traits of the aphid appear to drive the symbiosis (Shingleton et al. 2005). Associations with ants are also evolutionarily labile among plants. Even myrmecophytes (i.e., plants with domatia and food-bodies to house and feed persistent, resident ant colonies) can lose their association with ants. For example, the detailed molecular phylogenetic analysis of Blattner et al. (2001) suggests that nonmyrmecophyte species of *Macaranga* evolved at least once from congeneric myrmecophytes.

One might anticipate that the evolution to independence is restricted to organisms for which the symbiosis is facultative and that, where symbiosis is obligate, symbiosis collapse results in extinction. This reasonable expectation is not borne out by the data. The symbiosis with *Wolbachia* is obligate for filarial nematode hosts, but the phylogeny of these nematodes includes taxa that have abandoned the symbiosis nested within the symbiotic clade (Casirhagi et al. 2004) (figure 2-8). There are comparable instances among the insects. Most leaf-hoppers are plant sap feeders and dependent on the obligately vertically transmitted bacteria *Sulcia muelleri* and *Baumannia cicadellinicola*, but the typhlocybine leaf-hoppers have reverted to feeding on whole plant cells and have secondarily lost the symbiosis (Buchner 1965). The cockroaches and termites are now widely accepted as sister groups (Lo et al. 2000), and the symbiosis with the intracellular *Blattabacterium cuenoti* is universal in cockroaches and present in a single, basal species of termite, *Mastotermes darwiniensis*. Perhaps the obligacy of symbiosis for these animal hosts is shaped strongly by diet or is otherwise easily reversible by a few mutations. An alternative possibility that cannot be excluded is that the common ancestor of the symbiotic and nonsymbiotic animals did not require the symbiosis, and that obligacy has evolved subsequently in the lineage giving rise to the extant symbiotic hosts.

2.5 Constraints on the Evolution of Symbioses

The main focus of this chapter has been the diversity of routes by which the symbiotic habit has evolved. Nevertheless, the evolutionary opportunities afforded by symbiosis are not limitless. In particular, there are two intriguing large-scale phylogenetic constraints on the distribution of symbioses: the distribution of symbiosis with eukaryotes among the bacteria, and of intracellular symbioses among animals.

Of the two domains of bacteria, the Eubacteria and Archaea, only the Eubacteria are eukaryophilic, including various members that are pathogens and symbionts of eukaryotes. There are no known archaeans

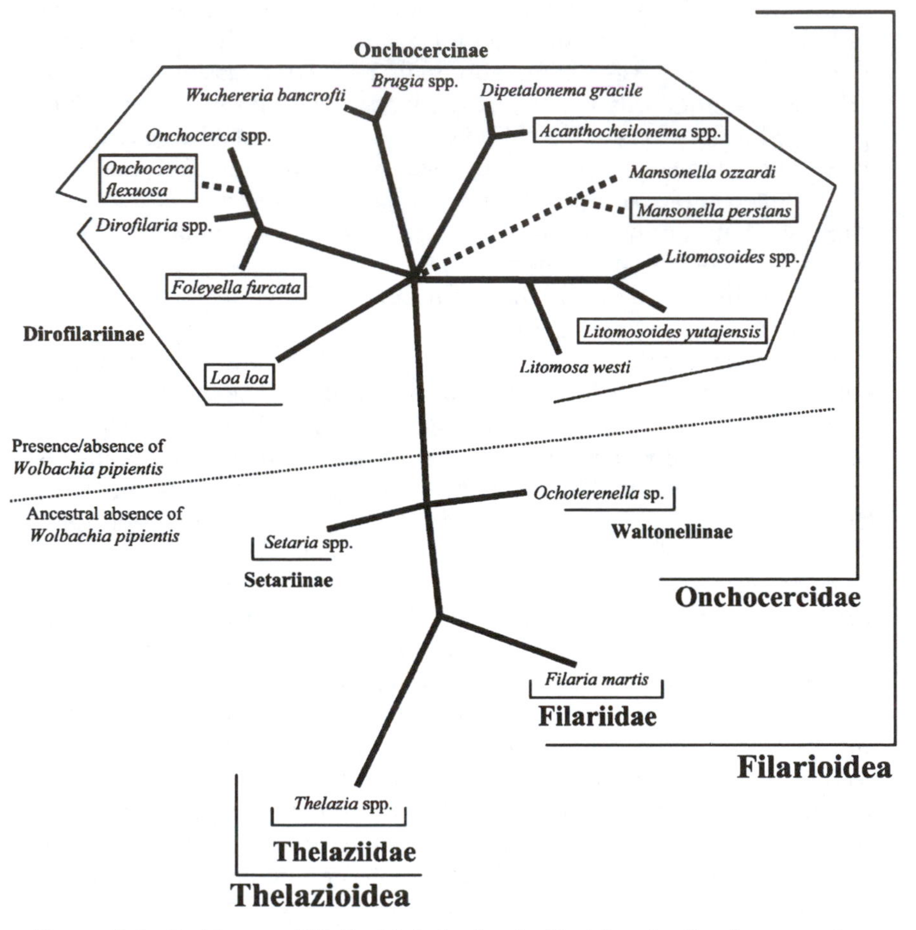

Figure 2-8 Incidence of *Wolbachia* infection in filarial and related nematodes. The symbiosis is envisaged to have evolved in the lineage leading to subfamily Onchocercinae (above the oblique line) and subsequently lost in the several taxa outlined in boxes. The dashed branches indicate species for which the taxonomic affiliation was not checked by gene sequence analysis. [Reproduced from *International Journal of Parasitology*, figure 4 of Casirhagi et al. (2004) with permission from Elsevier]

that are human pathogens [for a brief discussion of this topic, see Reeve (1999)], and archaean symbionts of eukaryotes are restricted to methanogens in some anaerobic protists (see chapter 3, section 3.2.3) and in the guts of many animals. Even so, the colon of humans bears just one abundant archaean species, *Methanobrevibacter smithii*, and many hundreds of species of Eubacteria (Eckburg et al. 2005). The reason why the Archaea generally do not engage with eukaryotes is obscure.

Among animals, intracellular symbionts are widespread among invertebrates but they have never been reported in vertebrates. It is tempting to attribute this difference between vertebrates and invertebrates to the evolution of cell-mediated adaptive immune system in the early vertebrates. Specifically, all cells in the vertebrate body present peptide fragments of intracellular proteins on their cell surface in association with class I major histocompatibility (MHC) proteins. If circulating $CD8^+$ cytotoxic T lymphocytes recognize any MHC-associated peptides as non-self, they trigger apoptosis (programmed death) of the cell, killing all the microorganisms it might contain. Crucially, the specificity of the T lymphocytes is generated by random somatic recombination in each animal, and only those epitopes that are displayed early in development when the immune system is being educated to tolerate self will fail to trigger cell death. Because lymphocyte specificity is both random and somatic (Boehm 2006), there is no opportunity for selection on the immune system to tolerate specific intracellular symbionts acquired by vertebrates with a mature immune system. The defense has been subverted by various pathogens, including taxa such as *Mycobacterium tuberculosis* which can persist in host cells for long periods without any deleterious effect on the host (Bhavsar et al. 2007). This leaves as an open question why no beneficial symbiont is known to have taken a similar evolutionary route. Perhaps intracellular symbioses in vertebrates do occur but are rare and have yet to be identified. For comparison, intracellular symbioses in bacteria were long believed not to exist because bacteria lack the capacity to internalize large particles by endocytosis; but the microbial partner in symbiosis with one group of insects, the mealybugs, is now known to comprise a γ-proteobacterium within a β-proteobacterium (Von Dohlen et al. 2001).

2.6 Résumé

The most striking aspect of the evolutionary origins and fates of organisms in symbioses is their diversity. Natural selection can make and break symbioses in multiple different ways. The immediate ancestors of symbiotic organisms can be free-living or participants in antagonistic associations or in symbioses with different partners. Similarly, some organisms in symbioses appear to be specialized to that lifestyle, but there is also phylogenetic evidence for various nonsymbiotic organisms and, less commonly parasites, with a symbiotic ancestry.

Most research on the evolution of symbiosis has focused on antagonistic origins. Amelioration (reduced virulence) and addiction (dependence without benefit), together with mode of transmission, have all been proposed as important processes contributing to the transition from antagonism to mutually beneficial relationships. The transition can occur rapidly (e.g., less than two decades for *Wolbachia* in natural *Drosophila* populations; see section 2.2.3) and be mediated by just one to a few genes (section 2.2.1). One implication is that organisms involved in parasitic or other antagonistic relationships are predicted to have many features in common with related organisms in symbioses. These features include the capacity to engage with a partner, especially with its defense systems, signaling pathways and nutrient allocation patterns. The nature of these traits is explored in chapter 4.

Nevertheless, the emphasis in the literature on symbiotic organisms that are related to organisms in antagonistic associations can be misleading. Very many symbiotic organisms have not evolved from parasites or pathogens. In particular, there is ample evidence, especially from microbial consortia, that various free-living organisms with complementary capabilities but no prior history of symbiosis can form mutually beneficial associations under certain conditions. Additionally, phylogenetic data indicate that at least one of the partners in some associations has had prior history of symbiosis with different organisms (section 2.3). Thus, the traits of an organism required for a particular symbiosis are not necessarily unique to that association, but can mediate a symbiosis with different taxa. The molecular basis of this capacity has been established in outline for one system, the propensity of plants to form root symbioses, and this is discussed in chapter 4.

Whatever their evolutionary history, the organisms in symbioses are expected to encounter conflict. As explained in chapter 1 (section 1.2), symbioses evolve and persist in the context of antagonistic interactions with predators, competitors, etc., and abiotic stresses, such as low nutrient availability. In addition, some instances of partner exchange probably involved antagonistic interactions between resident and incoming taxa, and partners in symbioses can come into conflict over access to common resources and the supply of costly services. Indeed, conflict and its resolution play a central role in the evolution and persistence of all symbiosis, and this is the topic of chapter 3.

One further issue is that the predisposition for the symbiotic habit is far from universal. This can be illustrated at multiple phylogenetic scales. The Eubacteria are far more disposed than the Archaea to symbiosis; intracellular symbiotic bacteria have evolved multiple times among invertebrates, especially insects, but are unknown among vertebrates; and the propensity of plant roots to for mycorrhizas is not displayed by the several plant families that are resistant to infection by mycorrhizal fungi. Trade-offs with other traits may limit the incidence of symbiosis in some instances. For example, the apparent absence of intracellular symbioses in vertebrates may be an opportunity cost of the adaptive immune system (section 2.5). More generally, the uneven phylogenetic distribution of symbioses indicates that symbiosis is not a free-for-all. Symbiotic organisms are selective in the partners with which they associate, discriminating against organisms that lack certain traits; and this selectivity is crucial to the formation of symbiosis at the scale of the individual organism, as well as to the pattern of evolutionary origins and phylogenetic distribution of symbioses. I address the processes underlying the selectivity of symbioses in chapter 4.

CHAPTER 3

Conflict and Conflict Resolution

THE BRITISH NOVELIST Ian McEwan inadvertently described the core problem of symbiosis: "This is our . . . conflict—what to give to others, and what to keep for yourself" (*Enduring Love* 1997). Conflict is inherent to the reciprocal exchange of benefits that underpin symbiosis. As considered in chapter 1 (section 1.2), reciprocity assumes that the cost of providing a service is lower than the benefit of receiving the reciprocated service from the partner. This scenario is illustrated in figure 1-1a, reproduced as figure 3-1a here. There is an obvious problem. Why not eliminate the cost by accepting the service from the partner while providing nothing in return? An organism obtains greater net benefit by not providing a service, as illustrated in figure 3-1b. Organism-X in figure 3-1b is a cheater because it derives enhanced fitness by being ineffective as a partner in a symbiosis. Against this backdrop, how can various symbioses founded on reciprocal exchange persist over an organism's lifetime and over millions of years of evolutionary time?

In this chapter, I focus on the nature and incidence of conflict in symbiosis (sections 3.2 and 3.3) and the routes by which conflict is suppressed or resolved (sections 3.4 and 3.5). In the final section, I consider in some detail one route by which conflict has been resolved that results in the evolution of symbiosis-derived organelles (section 3.6).

3.2 THE SOURCES OF CONFLICT

3.2.1 Generic Models of Conflict in Symbioses

Our understanding of conflict in symbiosis has been influenced very strongly by a single model developed by Dreschler and Flood in 1950 and subsequently dubbed the Prisoner's Dilemma. Indeed, figure 3-1 describes the Prisoner's Dilemma model in a format intuitive for symbiosis.

The Prisoner's Dilemma model has a grave limitation: that the levels at which the costs and benefits are set (i.e., the payoff structure) are entirely arbitrary. If the payoff is changed equally for the two partners, the outcome changes [this is explored in detail by Kollock (1998)]. A further limitation for many real symbioses, especially those involving

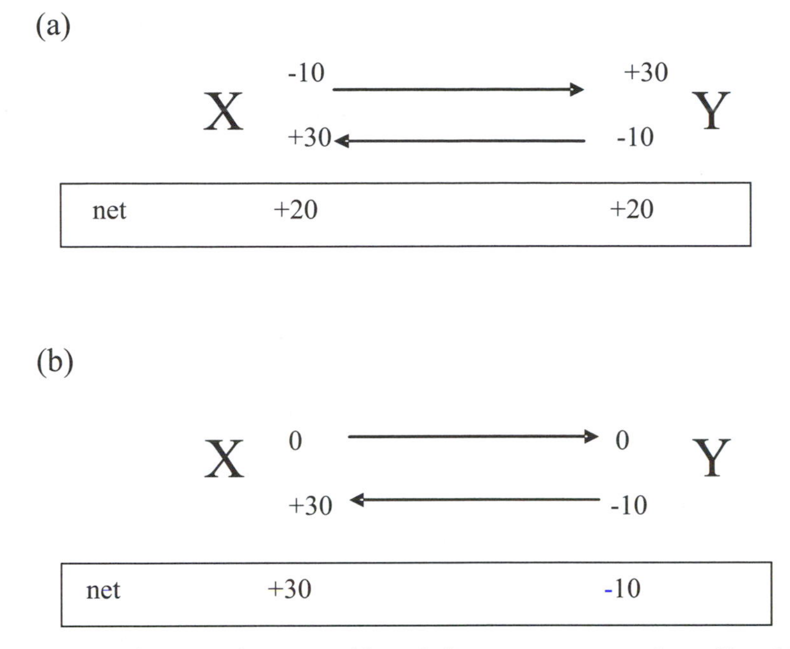

Figure 3-1 The cost of reciprocal benefit between two organisms (X and Y). (a) The cost of reciprocal benefit: each organism provides a service at cost of 10 arbitrary units and receives a benefit of 30 units, yielding net benefit of 20 units for each organism (figure reproduced from figure 1-1a). (b) The temptation to cheat: organism X does not provide the service and obtains increased net benefit (30 units) to the detriment of organism Y, which derives a net cost of 10 units. [Modified from figure 1 of Douglas (2008)]

a single large host and many small symbionts, is that the payoff can differ between the two partners, as I discuss later in this chapter (section 3.4.2). Clearly, we should not treat the Prisoner's Dilemma as an accurate representation of conflict in symbiosis. Even so, the model is important because further research has revealed one route by which conflict arising from costly services can be resolved. Although an individual behaving cooperatively is at a disadvantage in a single interaction (as in figure 3-1b), it can be very successful when the interaction is repeated many times. This counterintuitive conclusion emerged from a series of computer tournaments in which different strategies were tested against each other (Axelrod 1984). One strategy that performed particularly well was Tit-for-Tat, which followed the simple rule to do whatever the partner did in the previous interaction. Tit-for-Tit is nice (it starts by cooperating), provokable (it responds to cheating by the partner in one interaction by cheating in the next interaction), and forgiving (it always cooperates where the partner cooperates in the previous interaction). This strategy promotes cooperation in partners and performs well by rarely losing (unlike the losing partner Y in figure 3-1b).

A defining characteristic of symbioses is their persistence, meaning that the partners interact repeatedly (chapter 1, section 1.3). The computer simulations show that the nice-provokable-forgiving traits of the Tit-for-Tat strategy can drive the equilibrium position of the interaction towards the condition in figure 3-1a. In other words, symbioses founded on the Prisoner's Dilemma model of reciprocity are feasible *in silico*.

Another refinement to the Prisoner's Dilemma increases its realism for symbioses. This is the opportunity for a player to opt out of interactions with partners that cheat. Players with cooperative strategies (such as Tit-for-Tat) tend to avoid cheaters, so further increasing the probability of sustained cooperation. We can take this scenario one step further. If repeated interactions with one partner reveal it to be a cooperator, then a player may have a vested interest in the continued well-being of that partner. In biological terms, the fitness of the player is promoted by the fitness of its partner, and there is selective overlap between the two organisms. The greater the overlap, the more they are bound to cooperate. I return to this issue in section 3.4.1.

There is one further issue to address: mutualistic interactions founded on single interactions, such as the brief interaction between many plants and their pollinators. These relationships are predicted to be particularly vulnerable to cheating (figure 3-1b). The difficulty can be resolved if each partner can be assured that the partner will reciprocate; without that assurance, it is better to cheat. The best way to obtain such assurance is to be able to identify cooperative players in advance of an interaction. This creates the selection pressure for the evolution of honest signals, i.e. signals that cannot readily be faked and reliably advertize cooperation. The reader is referred to the excellent reviews of the Kollock (1998) and Sachs et al. (2004), where these nonsymbiotic interactions are considered.

3.2.2 The Costs of Symbiosis

The Prisoner's Dilemma and other models exploring reciprocal benefit presuppose that it is inherently costly to provide a benefit to partners. Their biological relevance depends on the validity of this assumption.

The costs are self-evident in some relationships, notably the associations between plants and pollinating insects which are also seed predators. Figs (trees of the genus *Ficus*) are pollinated exclusively by agaonid wasps. The many, tiny flowers of *Ficus* are massed onto the inner surface of a cavity, called the syconium. Pollen-bearing female wasps enter the syconium by a small opening, called the ostiole. The wasps both pollinate flowers in the syconium and lay eggs in some of the flower ovules, which are diverted from seed production to wasp production. The next generation of wasps emerges and mates, and the females leave the syconium via the ostiole. The details of *Ficus*-wasp interactions vary widely, with some *Ficus* species providing specialized neuter flowers for wasp oviposition and some wasp species actively harvesting pollen into specialized pollen pouches, but the core cost remains: that *Ficus* reproduction is dependent on pollinators that consume seeds.

It is only because *Ficus* species are specialized to their exploitative pollinators that we call these relationships mutualistic. (They are also sufficiently long-lasting to be called symbioses, see chapter 1, section 1.3.3.) This point is vividly illustrated by the relationship between the moth *Greya politella* and the saxifrage plant *Lithophragma parviflorum*. At some locations, the plant has access to bee and fly pollinators which do not consume seeds and, here, pollination by *G. politella* is disadvantageous through reduced seed production relative to that provided by the alternative pollinators. At other sites, where *G. politella* is the sole pollinator, its services are beneficial because the seed set is zero in the absence of pollination (Thompson and Cunningham 2002). This example illustrates how ecological context can play a crucial role in shaping the balance between the cost of providing a service (oviposition sites for the insect) and the benefit of receiving a service (pollination). The association is symbiotic at sites lacking alternative pollinators.

Various nonsymbiotic mutualisms founded on animal behavior are particularly amenable to investigation of the costs associated with providing a service. Here, I consider two elegant studies on ant–lycaenid caterpillar interactions and cleaner fish.

The caterpillars of many lycaenid butterflies secrete a sugary solution that attracts protective ants, as already considered in chapter 2 (section 2.2.5). This secretion is costly to produce. When caterpillars of *Jalmenas evagoras* were reared in the laboratory, they stopped producing the sugary secretion and grew to a significantly larger adult weight in the absence of ants than when ant-tended (Baylis and Pierce 1992). The size difference is important because it is a good predictor of egg production in females and mating success in males, indicating that it is costly for the caterpillars to be tended by ants. In the field, this cost is outweighed by the benefit of protection from predators: field populations of *J. evagoras* are completely eliminated by predators when ants are excluded.

The costs of being cleaned have been studied in the relationship between the territorial damselfish *Stegastes diencaeus* and its cleaning goby fish of the genus *Elacatinus* (Cheney and Coté 2001). The damselfish swim from their territories to a cleaning station, where their ectoparasites are removed by a goby. At the study site on Barbados, the damselfish territories varied widely in their distance from the nearest cleaning station. The fish with a cleaning station within or near to their territory spent more time being cleaned and bore fewer ectoparasites than those with territories at a distance from the cleaning station, and no damselfish traveled more than 2.2 m to be cleaned. Further behavioral experiments revealed two costs incurred when a damselfish leaves its territory to be cleaned: being attacked by other territorial fish, and the risk that the unattended territory is invaded by other fish. The greater the distance traveled and the longer the fish is away from its territory, the greater the cost.

In many instances, it is difficult to quantify the costs of providing a benefit because partners often retaliate against organisms from which they derive little benefit; i.e., it can be costly not to provide the service. (I discuss this further in section 3.4.2.) Modeling approaches can circumvent these difficulties. For example, Thomas et al. (2009) constructed a metabolic model of *Buchnera*, the symbiotic bacterium in aphids, and found that their symbiotic trait of essential amino acid release to the host is costly. The modeled bacteria grew 7–10% more slowly *in silico* when they were constrained to release these nutrients to the host at *in vivo* rates. With the increasing availability of detailed genomic and analytical information on the metabolic pathways of symbiotic organisms that provide nutrients or bioactive compounds to their partners, it should become increasingly feasible to use modeling techniques to assess the cost of symbiotic traits.

A twist to the analysis of cost in symbiosis is the increasing evidence that some interactions are cost-free, as is explored in the next section. (It is, of course, impossible to demonstrate that an interaction is completely cost-free and formally we should describe these interactions as without detectable cost.)

3.2.3 *Symbiotic Interactions Without Cost or Conflict*

Cost-free traits are displayed by an organism entirely for itself and access by other organisms is an incidental by-product. For this reason, cost-free symbiotic interactions are also known as by-product mutualism (Conner 1995). Members of a "selfish herd" display cost-free cooperation: by joining a group (herd, flock, school, etc.), each individual reduces the risk of predation for itself and all the other members of the group.

Cost-free beneficial interactions in symbioses are evident where two organisms have different elements of a common metabolic pathway.

It is particularly advantageous in low-oxygen environments because anaerobic metabolism yields little energy and requires high metabolic efficiency. A key metabolic problem faced by many anaerobes is that the fermentation of compounds such as short-chain fatty acids and alcohols requires a low partial pressure of hydrogen (pH_2); at high pH_2, the oxidation of these compounds requires the input of energy, i.e., the reaction is endergonic (figure 3-2a). Low pH_2 can be maintained by forming a symbiosis with hydrogen-scavenging bacteria, such as methanogens which use hydrogen as a substrate for ATP production (figure 3-2b). The transfer of hydrogen between the two species reduces the pH_2 in the first partner and provides a sustained source of hydrogen substrate for the second partner, and the benefit to each organism is incidental to the partner. The cost-free transfer of hydrogen is displayed in various consortia between bacteria and also between anaerobic protists, especially ciliates, and intracellular methanogens (Fenchel and Finlay 1995). An equivalent cost-free interaction occurs between sulfate-reducing bacteria, which require access to inorganic sulfate, and sulfide-oxidizing bacteria, for which the sulfate is a waste product. The partners of the sulfate-reducers are anoxygenic photosynthetic bacteria, such as green sulfur bacteria or purple sulfur bacteria, or alternatively colorless aerobic sulfur bacteria. Of course, the cohabitation of the aerobic sulfur bacteria and anaerobic sulfate-reducers is only possible under conditions with steep oxygen gradients, such as within microbial mats.

A beneficial interaction that has been assumed widely to be cost-free is the production of honeydew by ant-tended hemipteran insects. Honeydew is the sugar-rich egesta derived from the plant phloem sap on which these insects feed, and it is produced whether or not the insects are tended by ants. [This relationship differs from the ant-tending of lycaenids, in which the production of the sugar solution by the lycaenid is induced by the presence of ants and is undoubtedly costly, as described above (section 3.2.2).] Even so, there is evidence that some hemipterans might modify honeydew production to promote ant attendance despite an immediate physiological cost. The aphid *Tubercalatus quercicola* produces smaller droplets of honeydew at a higher rate and containing higher concentrations of amino acids and sugars when tended by ants than in isolation, and this correlates with depressed growth and reproduction of the ant-tended aphids (Yao and Akimoto 2002). It is thought that the aphids respond to the presence of ants by feeding more rapidly and assimilating fewer of the ingested nutrients. In this way, *T. quercicola* is investing in the relationship with ants to its own nutritional cost.

The differences between cost-free and costly beneficial interactions can also be illustrated by nitrogen recycling in symbioses between

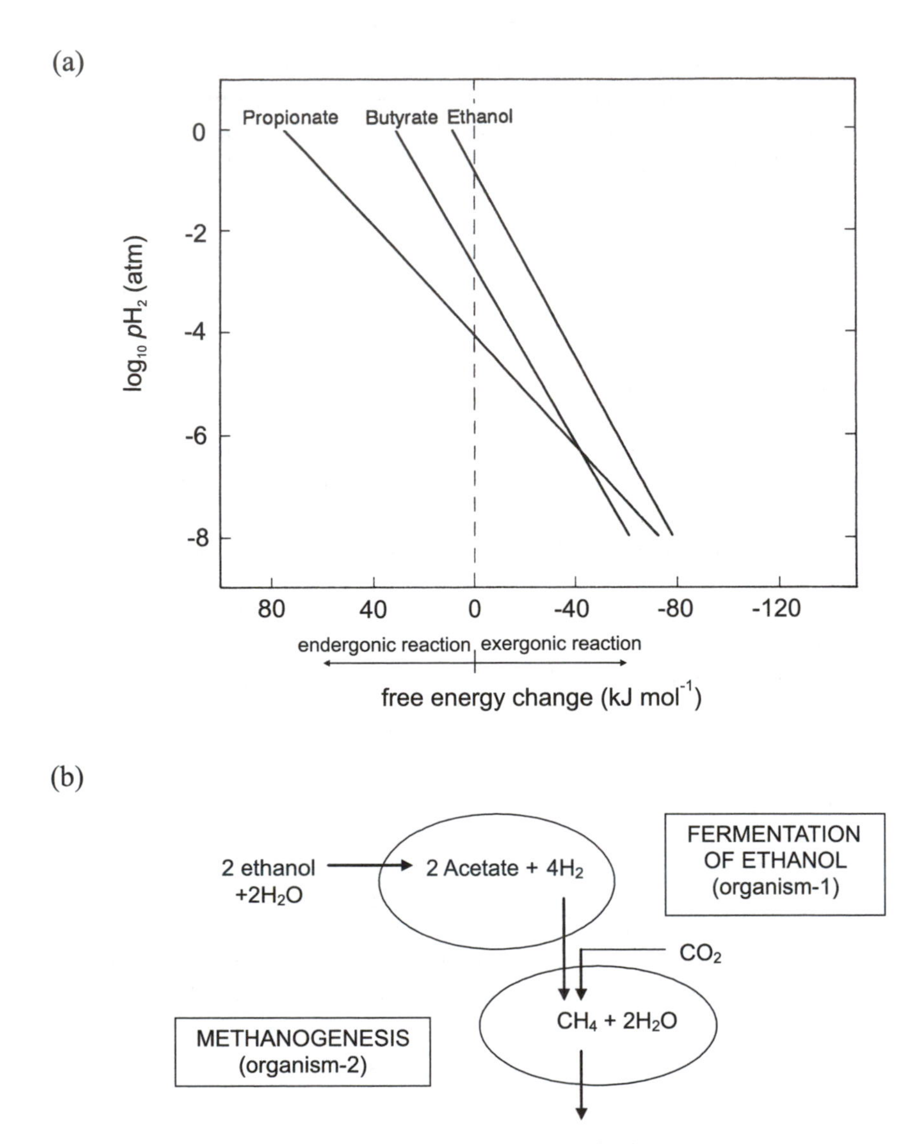

Figure 3-2 The metabolic basis of hydrogen transfer between species, a symbiotic interaction without cost. (a) The impact of partial pressure of hydrogen on the free energy change associated with the oxidation of three different organic substrates (propionate, butyrate, and ethanol). (b) The utilization of ethanol by a symbiosis between an anaerobic bacterium fermenting ethanol and a methanogen. [Redrawn from figures 2.24 and 2.23 of Fenchel and Finlay (1995) by permission of Oxford University Press]

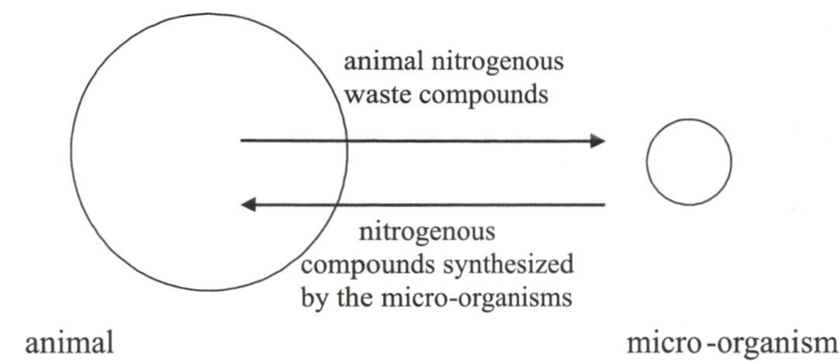

Figure 3-3 Nitrogen recycling in animal-microbial symbioses. The microorganisms transform nitrogenous waste products of the animal (ammonia, urea, etc.) into nitrogenous compounds valuable to animal metabolism, and these compounds are translocated back to the animal tissues.

animals and microorganisms. For the animal, nitrogen recycling is like alchemy—transforming not lead to gold, but toxic wastes to essential nutrients. It comprises two steps: a cost-free transfer from the animal to its microbial symbionts, and a costly transfer in the reverse direction (figure 3-3). I will take them in turn.

The cost-free step is the consumption of animal nitrogenous waste compounds by the symbiotic microorganisms. The animal benefits because the microbial symbionts act as a sink for potentially toxic nitrogenous waste compounds; and the symbionts benefit from access to a nitrogen source for growth. This interaction appears to be cost-free. There is no evidence that any animal produces more nitrogenous waste than is dictated by its own metabolism, just to feed its symbionts, or that the symbionts consume the host-derived nitrogen at rates higher than dictated by their own nitrogen requirements, just to reduce the load of nitrogenous wastes on the animal. The second step in nitrogen recycling is predicated on the symbionts being metabolically more versatile than the animal host, specifically that they can use the animal-derived nitrogen as a substrate to synthesize nitrogenous compounds which the animal cannot produce—and then to release these compounds back to the animal. Candidate compounds for release are essential amino acids (the amino acids that contribute to protein but with a carbon skeleton that cannot be synthesized by animals). This second step is costly to the symbionts because essential amino acid transfer to the host reduces the pool of these nutrients to support microbial growth. Nitrogen recycling has been invoked for associations with symbiotic algae, including the dinoflagellate *Symbiodinium* in corals and sea anemones, and prasinophyte algae *Tetraselmis* in the flatworm *Symsagittifera* (=*Convoluta*) *roscoffensis*, and also in insect symbioses with bacteria, especially in cockroaches and termites (Tanaka et al. 2006b; Douglas 1983; Mullins and Cochran 1975).

To summarize, the defining diffference between beneficial interactions that are cost-free and costly relates to their impact on the fitness of the organism conferring the benefit: costly traits reduce fitness, while cost-free traits are fitness-neutral. These traits differ in two further ways. The first concerns the response of an organism to separation from the symbiosis. An organism is predicted to continue to display a cost-free trait, since the trait is displayed entirely for the organism's own advantage, but not a costly trait, which is advantageous only in the context of a reciprocated benefit (see figure 3-1a). The second difference relates to the expectation that costly traits are symbiotic adaptations which have evolved and persist only in the context of the reciprocated benefit. Many nonsymbiotic taxa, including forms related to organisms in symbiosis, are expected to display a cost-free trait but not a costly trait. These predictions are illustrated by nitrogen recycling. As expected for cost-free traits, all animals, including symbiotic animals deprived of their microbial symbionts, produce nitrogenous waste (e.g., ammonia, urea, uric acid), and the capacity to utilize these compounds is widespread among microorganisms, including symbionts implicated in nitrogen recycling and brought into pure culture. By contrast, nonsymbiotic microorganisms do not display the large-scale selective release of specific compounds and the algal symbionts of animals are notorious for their cessation of nutrient release on isolation from the association; and these are the features expected for costly beneficial traits, such as the release of essential amino acids in nitrogen recycling.

I suspect that cost-free beneficial traits are important in many symbioses. They may play a central role in the evolutionary origin of symbioses from chance encounter, providing a context of mutual benefit within which costly traits can evolve (see chapter 2, section 2.4.1). They may also contribute to the persistence of symbioses, especially under conditions where costly traits are particularly costly. Cost-free traits are generally perceived to be uninteresting because they lack the evolutionary conflicts and dilemmas inherent in costly traits. This is probably why they tend to be neglected in symbiosis research, despite their potential importance.

3.3 Cheating Symbiotic Partners

Although computer simulations of interactions between two organisms suggest that it does not pay to cheat when the partners interact repeatedly (see section 3.2.1), cheating has been reported in a number of real symbioses. Before considering this in detail, it is important to reiterate exactly what is meant by the term cheater and the types of organism that cheat.

Cheaters are organisms that enhance their own fitness by reducing their costs through conferring little or no benefit on their partner. Two points follow. First, cost-free traits, such as the consumption of animal nitrogenous wastes by symbiotic microorganisms (figure 3-3), are not susceptible to cheating. Second, an organism which provides poor services to its partners is not necessarily a cheater—it could simply be ineffective, deriving no advantage from failing to benefit its partner. To illustrate, the nitrogen-fixing rhizobia vary widely in their capacity to promote the growth of their leguminous plant hosts. In one superbly detailed study (Burdon et al. 1999), the growth of seedlings of various Australian *Acacia* trees varied up to tenfold, depending on the rhizobial strain. The rhizobia that nodulated but supported little growth could have been cheaters or simply ineffective; and their ineffectiveness may have been general or strongly context dependent, varying with host genotype or developmental age, soil conditions, or some other environmental factor. As this example illustrates, it can be far from straightforward to identify cheaters in symbiosis.

Some organisms can behave as symbionts or cheaters, depending on environmental circumstance, including the identity of their partner. These can be described as opportunistic cheaters. Other cheaters are professional, comprising species that invariably exploit the association without providing benefit to the partners. A few professional cheaters are closely related to a partner in a symbiosis, and have probably evolved from the symbiotic partner; and many are distantly related and these are unlikely to have had any evolutionary history of involvement in the symbiosis that they exploit (although they could be derived from a symbiotic partner that has subsequently gone extinct). The distinction between opportunistic and professional cheaters can be illustrated by the nonsymbiotic cleaner relationships. The cleaner has the choice whether to cooperate by foraging for ectoparasites or to cheat by consuming healthy skin and flesh. The advantages of cheating to the cleaner are reduced foraging time and possibly higher-quality food, and the cost of a cheating cleaner to the client is persistent ectoparasites, often compounded by direct damage. The cleaner fish *Labroides dimidiatus* cheats opportunistically. Although it is generally an effective cleaner, it can also take a bite of flesh from its client while foraging for ectoparasites; and this trait appears not to be accidental because, remarkably, cheating is reduced in cleaners that can see other potential clients watching them (Bshary and Grutter 2006). The relationship between *L. dimidiatus* and its client fish is exploited by a professional cheater, the saber-toothed blenny *Aspidontus taeniatus* (Wickler 1966). The two species are very similar in size and coloration. Unlike most blennies, which have a blunt head with protruding eyes, *A. taeniatus* has a tapering head with lateral eyes, like *L. dimidiatus*. Furthermore, the coloration and markings of *A. taeniatus* closely match the local *L. dimidatus* populations, which vary with geographical location. *A. taeniatus* is not, however, a perfect mimic and client fish learn to avoid the cheater; young, inexperienced fish are the usual victims.

The incidence of professional cheaters has been studied intensively among seed predator pollinators. *Yucca* plants are actively pollinated by *Tegeticula* moths. The female moth collects pollen from one flower and then flies to another flower at a different developmental stage, where she forces the pollen down the receptive stigma with her mouthparts and then deposits eggs via the ovipositor. Detailed morphological and molecular studies have revealed that *Tegeticula* comprises at least 16 species, two of which have secondarily abandoned pollination, and so made the transition from symbiont to cheater (Althoff et al. 2006). The cheating species would go extinct in the absence of the symbiotic species, which ensure that the host plant reproduces. Even so, the cheaters can substantially depress seed set of the *Yucca* plants. The cheaters in the association between between figs and fig wasps, differ from *Tegeticula-Yucca* in that they are mostly unrelated to the pollinator wasps (Cook and Rasplus 2003). The cheating habit has evolved multiple times, and a single fig species pollinated by just one or two species of agaonid wasp species can be exploited by up to 30 nonpollinating seed predators (Marussich and Machado 2007). Many of the cheaters have long ovipositors, enabling them to deposit eggs from the external surface of the syconium, and others gain entry to the syconium via the ostiole, in the same way as the pollinators.

Professional cheaters have been documented in a number of ant-plant associations. For example, ant-tended *Piper* plants are invaded by beetles of the genus *Phyllobaenus*, which eat both the ants and food-bodies, while offering no protection (Letourneau 1990). Some cheaters are very specific. The ant *Cataulacus mckeyi* exploits a single symbiosis, the association between *Petalomyrmex phylax* and the West African tree *Leonardoxa africana*. It can invade the symbiosis because its head fits through the entrance holes into the domatia of the plant, as is discussed further in chapter 4 (section 4.2.1). Unlike the mutualistic ant partner which actively patrols the plant, *C. mckeyi* is relatively inactive, traveling from its nest site only to feed on the ant-bodies and extrafloral nectaries (Gaume and McKey 1999).

Despite cheaters, symbioses do persist. I now turn to consider routes by which cheating is controlled and the conflict between symbiotic partners is resolved.

3.4 Routes to Conflict Resolution

3.4.1 *Mode of Transmission*

Just as conflict arises from a difference in selective interest between the partners of a symbiosis, conflict can be resolved by increasing the overlap in selective interest of the partners. Selective overlap is strongly influenced by the mode of transmission (Ewald 1994; see also chapter 2, section 2.2.3). An organism is selected to promote the well-being of its partner if its fitness and transmission opportunities depend on the partner. Selective overlap is particularly favored by vertical transmission; the fittest symbiont is the one that maximizes the number of host offspring (i.e., host fitness) because each offspring is a new habitat for colonization by the symbiont's propagules.

A crucial experiment that revealed the significance of vertical transmission in reducing conflict was conducted by Bull et al. (1991) on a bacterial virus, the filamentous f1 phage. This virus is horizontally transmissible by release from bacterial cells, and it can also be transmitted vertically by being integrated into the bacterial chromosome and copied passively with each round of bacterial chromosomal replication and cell division. The experiment used two strains of f1 phage which differed in their cost to the host, i.e., the extent to which they depressed the growth rate of the host *E. coli* bacteria. They were grown together in competition on *E. coli*. When cultured under conditions that prevented horizontal transmission, the less costly viral strain increased in number relative to the more damaging strain, but when grown under conditions that permitted horizontal transmission this advantage was lost. These results demonstrate that vertical transmission selects for partners that impose low costs.

The relevance of these results to symbioses among organisms has been demonstrated for various associations. For example, the acquisition of vertical transmission by endophytic fungi in grasses has been central to the evolutionary transition from pathogens to symbionts that protect some grasses against herbivores, as discussed in chapter 2 (section 2.2.3). An experiment that manipulated the mode of transmission of symbiotic algae *Symbiodinium* in the jellyfish *Cassiopeia xamachana* also provides compelling evidence for the importance of transmission mode (Sachs and Wilcox 2006). The starting material for the experiment was an alga-free culture of genetically identical hosts and pooled culture of algal symbionts derived from symbiotic *C. xamachana*. The algal culture was likely to include multiple genotypes that varied in their quality as symbionts. The experiment comprised two regimes. In the vertical transmission regime, a cohort of hosts was infected with the algal culture and then maintained through two asexual generations with

TABLE 3-1
Impact of Mode of Transmission on the Symbiotic Traits of the Alga *Symbiodinium* in the Jellyfish *Cassiopeia xamachana*

Transmission mode	Host growth rate (mg day^{-1})	Host reproductive rate (number of buds day^{-1})	Algal expulsion rate (number of cells h^{-1})
Horizontal	0.041 ± 0.005	1.44 ± 0.101	997 ± 227
Vertical	0.090 ± 0.004	2.24 ± 0.163	353 ± 88
	$P < 0.001$	$P < 0.05$	$P < 0.05$

Source: Table 1 of Sachs and Wilcox (2006) with mean and estimated s.e. shown.

vertical transmission. This regime is predicted to select for algae which promoted host fitness. In the horizontal transmission regime, algae expelled from the first cohort of infected hosts were administered to a new cohort of alga-free hosts over two consecutive generations. This regime is expected to select for infective algae. The symbiotic traits of the algae subjected to the two selection regimes were different. The vertically transmitted algae supported higher host growth and reproductive rates and released fewer algal cells available for horizontal transmission (table 3-1). These results are fully consistent with the analysis of f1 phage in *E. coli*: that vertical transmission can select for traits beneficial to the host.

Obligate vertical transmission (i.e., viable propagules are not shed and are unable to survive either in isolation or by infecting a different partner) particularly promotes selective overlap. This is because the organism has no opportunity to escape from the consequences of any deleterious impacts it might have on its partner. As a counterexample, the pathogenic HIV virus is both vertically and horizontally transmitted. Insight into the molecular mechanisms underlying the evolution of obligate vertical transmission comes from studies of the bacterium *Rickettsia peacockii*, a symbiont of the tick *Dermacentor andersoni*. *R. peacockii* is closely related to the pathogen *R. rickettsii*, which causes Rocky Mountain spotted fever. *R. rickettsii* and various other rickettsias have complex lifecycles involving the regular or occasional horizontal transmission via tick saliva to mammals on which the ticks feed. The horizontal transmission of *R. rickettsii* depends on its capacity to migrate between host cells. The rickettsia gene *rickA* recruits host cell actin to form long filaments known as actin tails attached to the bacterial cell. *R. peacockii* lacks the capacity to induce actin tails because the coding sequence of the *rickA* gene is disrupted by an insertion sequence (Simser et al. 2005). As a result of this single genetic lesion, horizontal transmission is blocked. The genetic basis of the mode of transmission in other symbioses remains to be established; and it is likely to be complex in the many associations, especially the various mycetocyte symbioses of insects, where vertical transmission is integrated into complex reproductive and developmental programs of the host (Buchner 1965).

Obligate vertical transmission is not, however, the panacea for conflict resolution in symbiosis. Conflict can occur in vertically transmitted systems, as I discuss in section 3.5, and the incidence of vertical transmission can be constrained by other traits of the symbiotic partners. Two important factors likely to limit the incidence of vertical transmission can be summarized as cost and anatomy.

It is costly to a symbiotic organism to house and nourish the partner during stages in the lifecycle or under environmental conditions where it derives no benefit from the association. Plant seedlings raised in isolation commonly perform poorly when infected with mycorrhizal fungi because they lack a fully developed photosynthetic capability required to meet the high fungal demand for carbon (Smith and Read 2007). The mycetocyte symbionts of insects are vertically transmitted and often proliferate rapidly after transfer to the egg (Buchner 1965), presumably utilizing nutritional reserves that could otherwise be allocated to the embryo. The second constraint on the incidence of vertical transmission, host anatomy, relates to ease of access to host reproductive organs. Among plants, vertical transmission is the norm for many shoot-borne symbionts, including bacteria in the leaf nodules of *Ardisia* species and cyanobacteria in leaflets of the water fern *Azolla*, but root symbionts, including the mycorrhizal fungi and most rhizobia, have no access to the flowers and seeds and are invariably horizontally transmitted. In animals, the gut wall is a major barrier to microbial colonization of the body cavity and, ultimately, the gonads; and gut-borne microorganisms are generally horizontally transmitted, while microorganisms in the body cavity or internal organs, e.g., mycetocyte symbionts of insects, are vertically transmitted.

Occasionally, these constraints on the evolution of vertical transmission have been overcome. In particular, the gut wall barrier is bypassed by the behavior of some animal hosts, resulting in the vertical transmission of some gut symbionts. For example, vertical transmission of the γ-proteobacterium *Ishikawaella* in the gut of plataspid insects (also known as stinkbugs) is achieved by the maternal deposition of fecal pellets bearing these bacteria, which are consumed by the larvae immediately on hatching from the egg; and obligately anaerobic microbial symbionts in the rumen of cattle and other ruminants are transmitted by regurgitation into the mouth of the mother, inclusion in saliva, and passage to the offspring via licking by the mother. These exceptions do not detract, however, from the generality that horizontal transmission is widespread, and conflict in symbiosis cannot be resolved exclusively by overlap of selective interest arising from vertical transmission. I now turn to other routes to resolve conflict.

3.4.2 Sanctions and Rewards

One route to resolve conflict is to reward partners that provide a service and punish those that do not. In part, this topic echoes the analysis of the Prisoner's Dilemma (see section 3.2.1), particularly the importance of immediate punishment of cheating partners for the success of the Tit-for-Tat strategy. Nevertheless, many real symbioses differ from the Prisoner's Dilemma model in that the balance of power or exploitation among the organisms participating in symbioses is not symmetrical. One symbiotic partner tends to be in control, enforcing the good behavior of the others. The controller is expected to be the partner with the least incentive to cheat. This could be the organism with the greatest selective interest in the relationship or the one which incurs the lowest costs. For many associations involving a single large host and many small organisms, the controller is expected to be the host. This is because the smaller partner generally has invested less in the current relationship; small organisms have higher intrinsic reproductive rates than larger organisms and so they can produce a greater number of propagules to disperse away from the association. Asymmetry in the incentive to cheat can also arise from differences in symbiotic costs. For example, the plant partners of mycorrhizas may tend to be the controller because they incur little cost. Although an estimated 20% of photosynthetically fixed carbon is allocated to the fungal partner, this does not represent a substantial cost for the plant because photosynthesis is commonly sink-limited (Bryla and Eissenstat 2005; Kiers and van der Heijden 2006).

For some symbioses, one potential control point is the supply of nutrients required for the growth and persistence of its partners. In particular, intracellular symbionts are dependent on the surrounding host cell for all their nutrients, and the profile of nutrients that they can access is determined by the transport properties of the bounding host membranes and metabolic traits of the host cell. Additionally, some symbiotic organisms are nutritionally fastidious and their requirements can only be met in the association. For example, the symbiotic bacteria in some insects have a very restricted capacity to synthesize amino acids and nucleotides (Zientz et al. 2004), and the arbuscular mycorrhizal fungi (AM fungi) have no capacity for saprotrophic growth in soils and are dependent on plant-derived sugars for carbon (Smith and Read 2007). This asymmetry provides a route by which the controlling

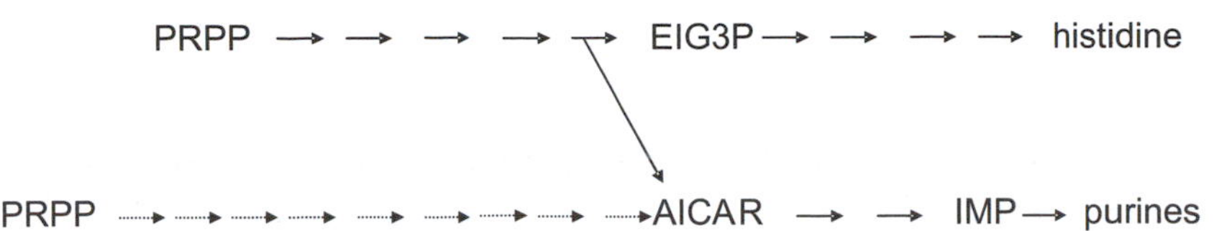

Figure 3-4 Metabolic coupling between the purine and histidine biosynthetic pathways in the symbiotic bacterium *Buchnera*. Sustained synthesis of purines depends on the supply of the metabolite AICAR from the histidine biosynthesis pathway (see text for details). Solid arrows: reactions present in both *Buchnera* and *Escherichia coli*; dashed arrows: reactions present in *Escherichia coli* only. The metabolites are PRPP: 5-phospho-alpha-D-ribose 1-diphosphate, EIG3P: D-erythro-1-(imidazol-4-yl)glycerol 3-phosphate; AICAR: 5-amino-1-(5-phospho-D-ribosyl)imidazole-4-carboxamide; IMP: inosine monophosphate. [Redrawn from figure 3b of Thomas et al. (2009)]

partner can reward partners that provide benefits and punish cheaters. Fitter (2006) has specifically proposed that the plant rewards AM fungi that provide phosphate by supplying them with photosynthetic carbon; the more phosphate provided, the more carbon allocated. This scheme is very plausible because plants are known to increase the allocation of photosynthate to portions of roots that encounter phosphate-rich patches in the soil.

Nutritional control over a symbiotic partner has also been proposed for the vertically transmitted bacterium *Buchnera* in aphids (Thomas et al. 2009). The control relates to the metabolic pathways for the synthesis of purines (including ATP and nucleotides for DNA and RNA synthesis) and the amino acid histidine, which is transferred at high rates to the host. A metabolite known informally as AICAR is produced as a by-product of one reaction in histidine synthesis and consumed in purine synthesis, as shown in figure 3-4. In *E. coli* and many other bacteria, AICAR is a metabolic intermediate in the purine biosynthetic pathway but, in *Buchnera*, this pathway is truncated such that AICAR is not synthesized. As a result, sustained production of purines by *Buchnera* depends on the supply of AICAR from the histidine biosynthetic pathway. Quantitative analyses have revealed that *Buchnera* must produce large amounts of histidine to make sufficient purines for its core requirements for DNA, ATP, etc. As a result of the coupled purine and histidine biosynthetic pathways, any *Buchnera* cell that does not overproduce histidine incurs the immediate and unconditional sanction of purine deficiency.

Some control measures exclusively comprise sanctions against cheating and other ineffective partners. The *Yucca* plants abort their fruits when their seed predator pollinator *Tegeticula* moths deposit very many eggs (Pellmyr 2002); and cleaning fish which cheat on their clients are often chased away (Bshary and Grutter 2005). One of the best-studied instances of sanctions comes from an experimental study of the symbiosis of *Bradyrhizobium* with soybean plants. To force the rhizobia to be ineffective, Kiers et al. (2003) prevented them from fixing nitrogen by replacing air with the N_2-free atmosphere of argon and oxygen. Whether the experiment was conducted at the scale of the whole root, part of the root system, or individual root nodule, the response was a reduction in the numbers of rhizobia (figure 3-5a). Monitoring of the oxygen relations revealed reduced oxygen tensions in the central infected zone of the nodule where the rhizobia are located, and depressed oxygen permeability of the outer nodule tissues (figure 3-5b). These results suggest that legume plants can impose sanctions on rhizobia which fail to fix nitrogen and that the sanctions include decreasing the oxygen supply to the rhizobia.

At this point, I return to the Prisoner's Dilemma and its limitations, including the poor match between the structure of the model (see section 3.2.1) and real symbioses with asymmetric controls. The Prisoner's Dilemma metaphor relates to two coequal partners (prisoners) responding to an externally imposed payoff structure (their jailor). For many symbioses, one of the partners approximates to the jailor. An additional complexity that is not considered in either the Prisoner's Dilemma or most experimental studies to date is that imposing sanctions is likely to be costly to the host through the diversion of resources and potential self-damage. Where the association is not obligate for the host, the host is long-lived, or alternative partners are readily available, it may be advantageous to eliminate the symbiosis altogether and either switch to a nonsymbiotic state or initiate a new symbiosis. The rhizobial symbiosis is suppressed in legumes reared on high-nitrogen soils, where the benefit to the plant of bearing nitrogen-fixing symbionts is small; and bleaching of corals has been interpreted as a route to dispose of failing algal symbionts (Buddemeier and Fautin 1993). Long-lived hosts could, in principle, cycle repeatedly through formation of the symbiosis with cooperative partners, the accumulation of cheaters, purging of the symbiosis to eliminate the cheaters, and acquisition of a fresh set of cooperative partners. An equivalent cycling between cooperation maintained by costly punishment, defection, and opting out has also been modeled for interactions among humans (Hauert et al. 2007).

3.4.3 *Restricting the Opportunity to Cheat*

The incidence of cheating depends on opportunity, and opportunity can be minimized by the controling partner. This issue has not been explored widely, but the ant-plant relationships provide an excellent

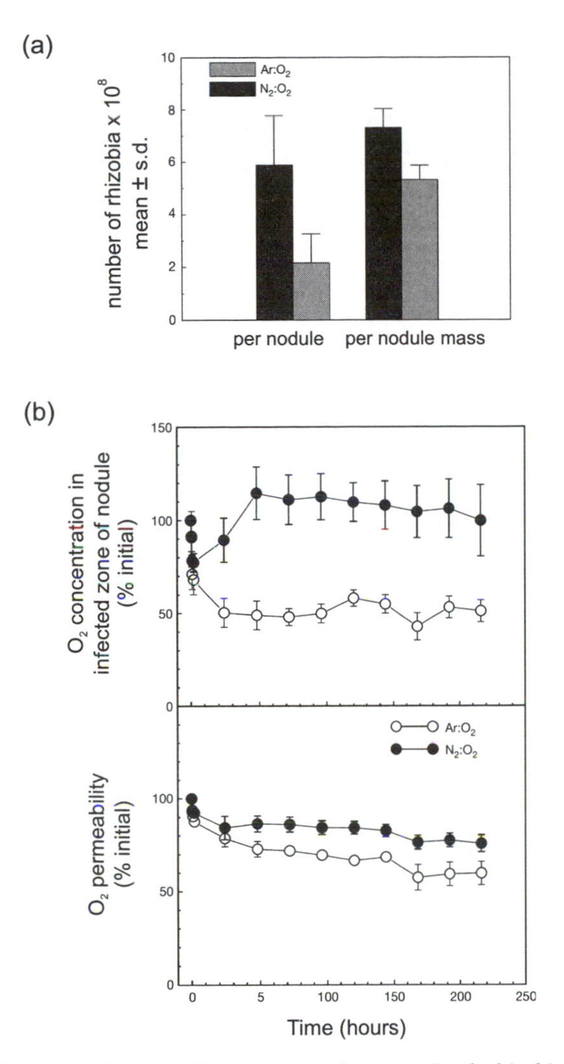

Figure 3-5 Sanctions imposed by soybean plants on *Bradyrhizobium* symbionts in which nitrogen fixation is prevented by exposure to nitrogen-free air (Ar: O_2, with nodules in N_2:O_2 air as controls). (a) Number of rhizobia in nodules. (b) Oxygen relations in nodules. [Redrawn from figures 2 and 3 of Kiers et al. (2003)]

example in relation to a conspicuous conflict: that the ants have no short-term selective interest in the reproductive output of their host plant. The ants may even benefit from depressed plant reproduction because flower and fruit production consumes plant resources that could otherwise be allocated to vegetative growth and production of ant-foods. This is probably why the *Allomerus* ants which protect the South American understory plant *Cordia nodosa* also systematically attack and destroy flower buds, preventing fruit production (Yu and Pierce 1998). Under some conditions, these ant colonies are essentially parasites, exploiting the increased growth and number of domatia of their castrated plant hosts.

Ant-mediated castration of plants is not, however, a common outcome of ant-plant conflict. As with many other plants, myrmecophytes (i.e., plants housing protective ant colonies) generally have physical or chemical barriers which reduce ant access to flowers. The domatia on myrmecophytes are also often sited at a distance from flowers. For example, *Hirtella myrmecophila* hosts *Allomerus* ants in pouches on young leaves, but these pouches are lost by abscission when the leaves mature. As a result, the flowers of *H. myrmecophila*, which are borne exclusively on branches bearing old leaves, are unlikely to be contacted by the potentially castrating ants (Izzo and Vasconcelos 2002). The opportunity of these symbionts to cheat is restricted by the structure of the host.

3.5 Conflict in Vertically Transmitted Symbioses

We have seen in section 3.4.1 how obligate vertical transmission tends to promote good behavior because of the broad overlap in reproductive interest between the host and symbiont partners. However, some conflict of interest remains in these associations, both between the host and symbiont, and among symbionts. I will consider these two types of conflict in turn.

3.5.1 Host-Symbiont Conflict over Transmission

A crucial proviso to the generalization that a host and its vertically transmitted symbionts have extensively overlapping selective interests is that the overlap relates strictly to the host individuals which mediate vertical transmission. For most sexually reproducing hosts with anisogamy (i.e., gametes of unequal size), the symbionts are transmitted via the egg, and therefore the shared selective interest is between the symbionts and female hosts; for a maternally inherited symbiont, a male host is an evolutionary dead end. Host-symbiont conflict ensues because the symbionts favor female host offspring but, under most conditions, the host favors a 1:1 sex ratio.

In most symbioses, the host sex determination mechanism is not susceptible to manipulation by the symbionts, but vertically transmitted

TABLE 3-2
Impact of a Bacterium Causing Cytoplasmic Incompatibility on the Incidence of Compatible (√) and Incompatible (×) Crosses between Hosts

Female host	Male host	
	Uninfected	Infected
Uninfected	√	×
Infected	√	√

microorganisms that alter host reproduction have been widely studied in insects and other arthropods, where they appear to be particularly prevalent. Enhanced transmission is achieved by manipulating host reproduction in three distinct ways (Charlet et al. 2003):

1. Parthenogenetic reproduction is induced in the host, to give twice the number of female offspring relative to uninfected hosts. Microbial-mediated induction of parthenogenesis appears to be restricted to host taxa with haplodiploid sex determination [i.e., where fertilized (diploid) eggs develop as females and unfertilized (haploid) eggs develop as males in uninfected hosts], and most examples of microbial-mediated parthenogenesis are in insects of the order Hymenoptera, especially wasps.
2. Male hosts are feminized, so doubling the number of female offspring, the same consequence as from microbial-mediated parthenogenesis (above). The principal taxa susceptible to feminization by microorganisms are Crustacea, specifically isopods (woodlice) and amphipods, but feminization of insects (e.g., the moth *Eurema hecabe*) has also been reported.
3. Uninfected eggs are killed by a factor associated with the sperm from infected hosts, but crosses between infected males and females, and between uninfected males and infected females, are fertile, as illustrated in table 3.2. This mode of host reproductive manipulation is usually known as cytoplasmic incompatibility. By killing uninfected eggs, the frequency of infected females in the population is increased, often to very high levels or fixation.

In addition, some vertically transmitted microorganisms preferentially kill male hosts. This sex-specific virulence does not result in an increase in the number of female offspring produced and, because the function of maternally inherited microorganisms is not under direct selection pressure in male hosts, male-killing can evolve without being advantageous to the microorganism. In other words, male-killing in a vertically transmitted microorganism is not evidence per se of host-symbiont conflict over transmission. However, if infected females benefit from the death of their brothers through reduced chance of inbreeding or depressed intersib competition, this trait can be to the selective advantage of the microorganism, and consequently spread through the host population (Hurst and Jiggins 2000).

Perhaps the most remarkable fact about the conflict over host sex ratio between hosts and their vertically transmitted symbionts is that it is not widespread. Reproductive distortion is apparently restricted to arthropods and is mediated by very few microorganisms. Most known instances involve the α-proteobacterium *Wolbachia*, the "master manipulator of host reproduction" (Hunter et al. 2003). A few other bacteria have been implicated, including *Cardinium* (Bacteroidetes) which causes cytoplasmic incompatibility, feminization, and parthenogenesis induction in several host taxa (Hunter et al. 2003), microsporidian protists which feminize and kill males, and several other male-killers, e.g., *Spiroplasma*, flavobacteria (Hurst and Jiggins 2000).

It appears that the conflict over host sex ratio has generally been won by the host: most vertically transmitted symbionts do not distort the reproduction of their hosts. Similarly, mitochondria and plastids in most hosts are uniparentally inherited without distorting host sex ratio; mitochondrial-mediated cytoplasmic male sterility in various plants is an exception that is consistent with the existence of this conflict. The widespread incidence of vertically transmitted microorganisms may have selected for host sex determination mechanisms that are difficult to distort, with the implication that microbial infections might be one selection pressure promoting the great diversity and complexity of sex determination mechanisms among eukaryotes.

Although reproductive distortion is rare, some associations with vertically transmitted symbionts that do not manipulate host sex ratio have traits which can be interpreted as markers of past conflict. For example, the symbiont-free males in some aphid groups (Buchner 1965) may have evolved to protect males from male-killing by a vertically transmitted symbiont; and males of haplodiploid coccids are entirely haploid, apart from the diploid mycetocytes housing vertically transmitted symbionts (Brown 1965), and this could be a route to disguise their maleness from the symbionts.

To summarize, there is overwhelming evidence that certain vertically transmitted microorganisms with maternal inheritance manipulate the reproductive system of their hosts. Vertical transmission is not an absolute guarantor of microbial good behavior. Although microorganisms such as *Wolbachia* demonstrate the existence of conflict between hosts and maternally inherited symbionts, the conflict in most associations is resolved strictly in favor of the host.

3.5.2 *Intersymbiont Conflict over Transmission*

There is a second source of conflict in vertically transmitted systems: that most members of the symbiont population are not transmitted to the host offspring. For example, in the symbiosis between the bacteria *Buchnera* and the pea aphid *Acyrthosiphon pisum*, an estimated 2–20% of *Buchnera* cells are transmitted, leaving up to 98% of the *Buchnera* cells to die with the maternal host. It is in the interests of the symbionts to compete for transmission, but in the interests of the host to suppress intersymbiont competition, which could damage the host.

A common characteristic of associations with vertical transmission is that the transmission process is highly ordered spatially and temporally (Buchner 1965), suggesting that any conflict among symbiont cells is suppressed effectively by the host. The spatiotemporal controls over transmission can be illustrated by the association between the human body louse *Pediculus humanus* and its bacterial symbionts, housed in a single organ known as the stomach disk, ventral to the stomach (figure 3-6a). In females, symbionts are retained within the stomach disk at all times except during the final larval molt; and, in males, they are never released from the stomach disk. The released symbionts in female lice migrate rapidly to the female reproductive organs (figure 3-6a), where they become incorporated into the pedicel at the base of each of the two ovaries (figure 3-6b). As in other insects, the louse ovary comprises multiple ovarioles (egg tubes), each with a linear array of eggs, the most developed at the base, abutting the pedicel. The basal egg in each ovariole is inoculated with a small number of bacteria from the pedicel, and it is then ovulated; the next egg in the ovariole is then basal, inoculated, and ovulated; and so on, such that every ovulated egg bears its complement of symbionts. This process is important for what does not happen. None of the bacteria exit the stomach disk prematurely, thereby gaining preferential access to the ovaries; and there is no killing of competing bacterial cells in the ovary pedicel. We are currently ignorant of the molecular mechanisms by which hosts maintain the tight spatiotemporal controls over vertical transmission of their symbionts, and it is unclear how difficult it would be for symbionts to subvert these mechanisms.

In many systems, the symbionts may not be selected to evade host controls over vertical transmission. At first sight, this statement sounds improbable. Who would not fight for the last place in the only lifeboat on a sinking ship, when the alternative fate is death? One answer to that question is identical twins; or, more correctly, there is no selection pressure for identical twins to fight. Returning to vertically transmitted symbionts, intersymbiont conflict is diminished through kin selection

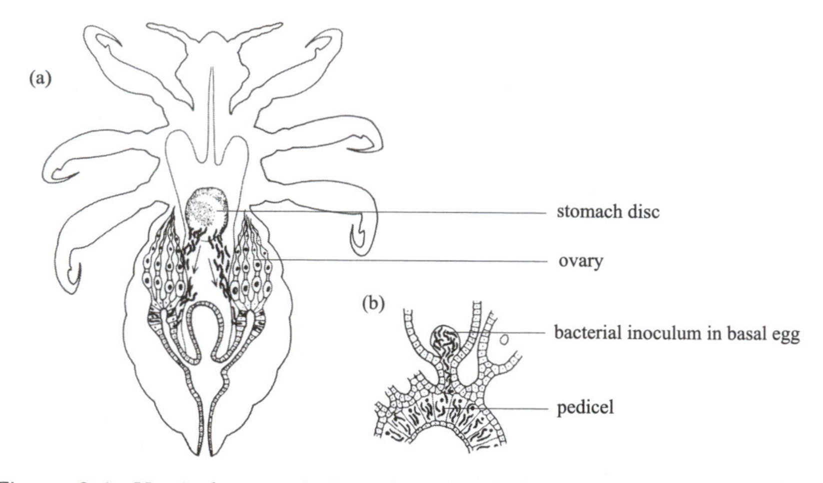

Figure 3-6 Vertical transmission of symbiotic bacteria in the human body louse *Pediculus humanus*. (a) Transmission of bacteria from stomach disk to ovaries of female insect. (b) Transfer of bacteria from pedicel at the base of each ovary to basal egg in one ovariole. See text for full description. [Redrawn from Eberle and McLean (1983) with permission from NRC Research Press]

where the symbionts are closely related, and collapses where symbionts are genetically identical. Genetic uniformity of symbionts is achieved by transmission of only a few symbionts from parent to each offspring over multiple host generations; in other words, the bacterial population experiences a very narrow bottleneck at each host generation. An equivalent argument accounts for the cooperation among the cells in the body of a multicellular organism, such that they do not compete to be gametes (Maynard Smith and Szathmary 1995). It is through kin selection that the counterintuitive conclusion is reached that competition for transmission is depressed by minimizing the number of transmitted symbionts.

The task of quantifying the size of the transmitted population of symbionts is technically challenging because the microbial symbionts tend to proliferate rapidly immediately after transmission (Buchner 1965). A rare dataset is the estimate of Mira and Moran (2002) that 850–8000 *Buchnera* cells are transmitted from an aphid host to each of her offspring. Further empirical studies are crucial to test this explanation for resolution of intersymbiont conflict over transmission. A further reason why this research is important is that small effective population size is widely invoked as a central element of explanations for the genomic traits of vertically transmitted symbionts. This is addressed in the next section.

3.5.3 *Symbiont Genomes and Genomic Decay*

Genome analyses have revealed three striking features in the genomes of intracellular bacteria, including symbionts that are exclusively vertically transmitted. Although their gene contents are far from uniform, their genomes tend to (1) be small (ca. 1 Mb or less), (2) be AT-rich (%GC <30%), and (3) evolve rapidly, with elevated substitution rates, including of nonsynonymous sites, i.e., nucleotide changes that lead to a change in the amino acid residues of the protein. These three genomic traits are associated. Genomes with high evolutionary rates generally have a high AT content because of a mutational bias toward AT, linked to the greater availability of the precursors ATP and TTP than GTP and CTP. Where the high rate of sequence evolution generates nonfunctional genes, those genes tend to be eliminated from bacterial genomes through a general deletional bias, believed to have evolved in bacteria to eliminate DNA parasites, such as transposons (Lawrence et al. 2001). Additional positive selection pressure for small, AT-rich genomes in intracellular bacteria may arise from the replication advantage of small genomes and the greater energetic cost of synthesizing GTP and CTP than ATP and TTP in resource-limited environments (Rocha and Danchin 2002).

In principle, these distinctive genomic characteristics of vertically transmitted symbionts can be explained by the retention of mutations that reduce the function of individual gene products. This, in turn, can be attributed largely to the tight bottlenecking of bacterial populations at transmission (see section 3.5.2), leading to the fixation of deleterious mutations and loss of gene function. Thus, resolution of intersymbiont conflict can have the "unintended" consequence of genomic deterioration of the symbiont.

To explain how the small effective population size of a symbiont can lead to genomic deterioration, let us consider a vertically transmitted bacterium, for example one of the ten million *Buchnera* cells in a single pea aphid. This *Buchnera* cell incurs a mildly deleterious mutation. Because *Buchnera* populations are bottlenecked at transmission (section 3.5.2), it is possible, by chance, for a fraction of the descendants of the aphid host to bear only descendants of the *Buchnera* mutant. Furthermore, each aphid invariably gains its complement from its mother, and there is no opportunity for genetically distinct bacteria to be acquired horizontally and enhance the genetic diversity of the population either directly or by genetic exchange with the resident population. Simply because the populations of vertically transmitted bacteria are small and predominantly asexual, multiple, mildly deleterious mutations can accumulate by genetic drift, causing genome-wide deterioration in an additive or ratchetlike manner. This process is referred to as Muller's ratchet (Muller 1964). Once the genes for DNA repair and recombination are compromised, genomic deterioration is accelerated.

The traits of a genome subject to decay can take one of two forms. In the first, the bacterial genome is ca. 1 Mb and has many mobile genetic elements or pseudogenes. For example, 14% of the genome of the *Wolbachia* wMel strain comprises repeated sequences and insertion sequence (IS) elements (Wu et al. 2004), while *Sodalis glossinidius* from tsetse flies has 972 pseudogenes taking up nearly half of the total genome (Toh et al. 2006). Several linked factors are likely to have contributed to the evolution of these traits: relaxed selection against mutations in genes not required for the symbiotic lifestyle, IS transposition into these genes, and small population size, which reduces the opportunity for elimination of pseudogenes and IS elements by recombination. These conditions of relaxed selection and small population size are not unique to symbiotic bacteria. They are realized by other bacterial lifestyles, including recently evolved pathogens (e.g., gene fragments account for ~50% of the *Mycobacterium leprae* genome), and even the repeatedly subcultured bacterium *Lactobacillus bulgaricus* (17% pseudogenes) used in traditional yoghurt-making (Cole et al. 2001; Van de Guchte et al. 2006).

The second type of genome subject to decay is compact (<0.8 Mb) and has few pseudogenes and no recognizable IS elements or other mobile genetic elements. For example, the genome of *Buchnera* in aphids is 0.45–0.66 Mb, of *Wigglesworthia* in tsetse fly is 0.7 Mb, and of *Carsonella* in psyllids is 0.16 Mb (Akman et al. 2002; Perez-Brocal et al. 2006; Nakabachi et al. 2006). It is tempting to treat the genomes rich in pseudogenes, IS elements, etc. as analogous to intermediates in the evolution of bacteria with very small genomes; that the nonfunctional sequences will gradually be deleted. This interpretation may, however, be a simplification. Some aspects of the genome organization of bacteria with very small genomes suggest that their early evolution included the elimination of one to several large genomic regions, possibly by recombination at repeated sequences (Klasson and Andersson 2003). In other words, single-gene inactivation events, as occurring in *Wolbachia* and *Sodalis* species, have undoubtedly contributed to—but may be insufficient to account fully for—the genome reduction leading to the very small genomes of *Buchnera*, *Carsonella*, etc.

Although genomic decay is recognized as an important determinant of the genomic traits of many vertically transmitted bacteria, this should not be interpreted to indicate that vertically transmitted symbionts are inherently transient, locked into an evolutionary trajectory that leads ineluctably to genomic meltdown. Various vertically transmitted symbioses have persisted for extended periods and are ecologically very successful. For example the aphid-*Buchnera* symbiosis is estimated to have evolved at least 160 million years ago (Moran et al. 1993), and the relationship between the bacterium *Sulcia* and other plant-sap-feeding insects (the cicadas, leaf-hoppers, plant-hoppers, etc.) is at least 260 million years old (Moran et al. 2005). Even so, the several instances of symbiont switching in insects (chapter 2, section 2.3.3) offer the tantalizing possibility of replacement of a resident decayed symbiont by a genetically vigorous alternative.

What may protect the genomes of vertically transmitted symbionts from genomic meltdown? Two factors are important: selection and compensation. Selection for function operates at the level of the entire symbiosis, and tends to protect symbiont genes with functions essential for persistence of the symbiosis. For example, *Buchnera* in pea aphids have retained many genes in the biosynthetic pathways for essential amino acids, which are released to the aphid and supplement the amino-acid-deficient aphid diet of plant phloem sap; but these bacteria have lost many other biosynthetic capabilities and depend on the aphid for various compounds, including most nonessential amino acids. Compensatory responses may occur in the symbiont or host. Many intracellular bacteria have high levels of the chaperonin protein GroEL, which generally directs the proper three-dimensional conformation of other proteins, and may ensure the stability and function of otherwise poorly functional proteins in the symbionts (Moran 1996). Host-mediated compensation includes the supply of metabolites to symbionts whose biosynthetic capabilities are compromised. As an example, *Buchnera* is totally dependent on the aphid host for various cofactors, including quinones in the electron transport chain (Shigenobu et al. 2000; Zientz et al. 2004).

Importantly, host-mediated compensation and selection for symbiont function are likely to have conflicting outcomes because compensatory responses of the host tend to reduce the selection pressure for symbiont function, and may even promote further genomic decay in the symbionts. In effect, host compensation can lead to the transfer of control over symbiont function from the symbiont to the host.

3.6 Genetic Assimilation: The Evolution of Symbiont-Derived Organelles

3.6.1 *Transfer of Symbiont Genes to the Host Nucleus*

In principle, there is a further route to compensate for the problem of genomic decay of vertically transmitted symbionts described in the previous section. This is the transfer of symbiont genes to the host nucleus, where they are subject to the nuclear regime of repair and recombination and, therefore, more able to respond to selection pressures for function (Blanchard and Lynch 2000). These genes have "escaped to sex." This selection pressure may be compounded by the advantage of small genome size in competition among the symbionts in a cell (symbionts with fewer genes can have smaller genomes and replicate faster than those with larger genomes). The genes which are functional after transfer are those that are transcribed and translated in the nucleocytoplasm with the cognate proteins targeted back to the symbiont (figures 3-7a and b). In this way, symbiont genes transferred to the host lead to the progressive genetic assimilation of the symbiont by the host.

There are just two definitive examples of vertically transmitted symbionts with functional genes transferred to the nucleus. These are an α-proteobacterium that evolved into mitochondria, and a cyanobacterium that evolved into the chloroplasts and other plastids (Kurland and Andersson 2000; see also chapter 1, section 1.4.3). It is estimated that more than 90% of the protein-coding genes with function exclusive to these organelles are located in the nucleus. In other words, the proteome of these organelles is more than ten times larger than their genome. It is from these observations that we have a working definition of a symbiont-derived organelle: its function depends on the products of genes that, during the evolutionary transition to an organelle, were transferred to the eukaryotic nucleus.

Genetic assimilation has profound evolutionary consequences. It leads to the expectation that sustained function of the symbiont-derived organelle is inescapably dependent on the host lineage in which it evolved, and the organelle cannot switch to a host lineage with which it has no shared evolutionary history. I explore this expectation further in section 3.6.5.

3.6.2 The Limits to Genetic Assimilation

In a host-symbiont relationship where symbiont genes can be sequestered in the host nucleus, one might expect gene transfer to the nucleus to go to completion until the genome of the symbiont/organelle genome is eliminated. By this reasoning, the final fate of a symbiont-derived organelle is genome-free. Such an organelle is genetically driven from the nucleus but never synthesized *de novo* by the nucleocytoplasm. It is the "ghost of a symbiont past."

The genetic assimilation of mitochondria and plastids is incomplete. The genomes of plant plastids are generally 100–150 kb and the protein-

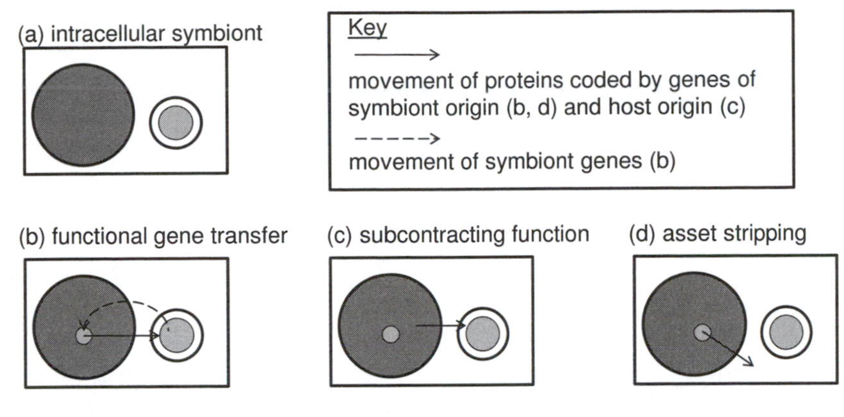

Figure 3-7 The genetic assimilation of intracellular symbionts. (a) A host eukaryotic cell (rectangle) with nucleus (large circle, with genome, (dark gray)) and cytoplasmic symbiont (small circle with genome, (light gray)). (b) The evolutionary transition from symbiont to organelle involves the transfer to the nucleus of symbiont genes, whose products are targeted back to the organelle; see section 3.6.1. (c) Subcontracting function, the targeting of host gene products to the symbiont-derived organelle; see section 3.6.3. (d) Asset stripping, the targeting of symbiont-derived genes to cell compartments other than the symbiont-derived organelle; see section 3.6.4.

coding capacity of mitochondria ranges from two genes (in *Plasmodium falciparium*) to 67 genes (in *Reclinomonas americana*) (Burger et al. 2003). The persistence of organelle genomes has been explained post hoc by two general processes. First, certain membrane proteins with extensive hydrophobic regions can be inserted into the correct organellar membrane in the correct conformation only when synthesized in the organelle (von Heijne 1986). Second, organellar function may depend on the regulation of expression of certain genes by the redox potential within the individual organelle, and redox control operating at the scale of the individual organelle is possible only if the gene is located in the organelle (Allen 2003). A third explanation relevant to some organelle lineages, notably mitochondria in animals, is that the organellar genetic code has diverged from the universal code, so precluding the translation of organellar genes on the cytoplasmic ribosomes (de Grey 2005). One consequence of the retention of organellar genes is genetic inefficiency. For example, the human mitochondrial genome has 23 genes coding for the translation machinery (ribosomal RNAs, tRNAs) required for expression of just 13 protein-coding genes.

Have any symbionts been completely assimilated by their hosts? There is evidence that one genome-free organelle, the hydrogenosome, has a symbiotic origin. Hydrogenosomes are double-membrane-bound organelles that mediate anaerobic ATP production via pyruvate ferredoxin oxidoreductase and hydrogenase. They have evolved independently from mitochondria in different groups of protists, including the trichomonads, amoeboflagellates, chytrid fungi, and ciliates (Embley et al. 2003). Evidence for a hydrogenosome-mitochondrion link is threefold. First, the N-terminal signal presequences targeting proteins synthesized in the cytoplasm to hydrogenosomes and mitochondria are experimentally interchangeable. For example, when the gene for the hydrogenosomal protein malic enzyme of the chytrid *Neocallimastix frontalis* was transferred to the yeast genome, the presequence mediated the efficient targeting of the protein to the yeast mitochondrion (van den Giezen et al. 1998). Second, several hydrogenosome proteins (e.g., the ATP/ADP translocator, ferredoxin, succinyl coenzyme A synthetase) in the trichomonad *Trichomonas vaginalis* are more closely related to mitochondrial than to bacterial sequences (Dyall et al. 2000). Finally, two orders of ciliates, the Heterotrichida and Scuticociliatida, include aerobic taxa with mitochondria and anaerobic taxa with hydrogenosomes that are structurally remarkably similar to mitochondria, suggesting strongly that the mitochondria have evolved into hydrogenosomes on multiple different occasions as their ciliate hosts switched from aerobic to anaerobic lifestyles (Fenchel and Finlay 1995). It appears that, as hydrogenosomes have evolved, they have been stripped of their DNA, while the related mitochondria which mediate oxidative phosphorylation have retained a genome. Perhaps the mitochondria (and plastids) are protected from genetic annihilation by their electron transport chains, through either redox control of gene expression or targeting of complex multisubunit proteins to the organelle membrane, as discussed above.

Genome reduction and loss are also involved in the evolutionary history of complex plastids. Complex plastids have evolved from eukaryotic algae, i.e., unicellular eukaryotes bearing plastids. In two algal groups, the cryptophytes and chlorarachniophytes, the nucleus of the eukaryotic algal symbiont has been retained as a reduced organelle known as a nucleomorph. As with the genomes of bacterial-derived organelles, these eukaryotic genomes are much reduced and simplified, just 551 kb (553 genes) in the cryptophyte *Guillardia theta* and 373 kb in the chlorarachniophyte *Bigelowiella* (Douglas et al. 2001). The nucleomorphs are the smallest known eukaryotic genomes and grotesquely inefficient. The *G. theta* nucleomorph, coding for just 30 proteins targeted to the plastid, is maintained by 434 nucleomorph gene products, including histones, tubulins, rRNAs, and DNA and RNA polymerases. Other complex plastids lack a nucleomorph, suggesting that the eukaryotic genome of the complex plastid has been completely assimilated, with its genes either eliminated or transferred to the secondary host nucleus. The only clue to the prior existence of these "symbionts past" is the complex membrane architecture of the plastids.

3.6.3 Subcontracting Function to Symbiont-Derived Organelles

In mitochondria and plastids, genetic assimilation by the transfer of genes to the nucleus is compounded by subcontracting of function, i.e., the targeting of gene products of host origin to the organelle, such that the function of the organelle is not limited to capacities coded by the genome of the microbial ancestor of the organelle (figure 3-7c). Most of the nuclear-encoded proteins recovered from the mitochondria and plastids are of host origin. For example, among the 423 mitochondrial proteins of the yeast *Saccharomyces cerevisiae* studied by Karlberg et al. (2000), about half were unambiguously of host origin, just 12% were identified positively as derived from the bacterial ancestor of the mitochondria, and the phylogenetic relationships of the remaining genes could not be distinguished with confidence.

By subcontracting function, symbiont-derived organelles can be exploited as compartments for functions unrelated to the biochemical capabilities of their ancestors. Thus, the biosynthesis of fatty acids in plants is mediated by enzymes exclusively of eukaryotic origin and localized to plastids, a convenient intracellular compartment (in animals, fatty acids are synthesized in the cytoplasm). Mitochondria are also used to store calcium ions and, in animals, they mediate fatty acid oxidation (in plants, most fatty acids are degraded in peroxisomes). The central role of mitochondria in programmed cell death may be a further example of subcontracted function; alternatively, as described in chapter 1 (section 1.4.3), this interaction may have evolved from an antagonistic interaction between the early eukaryotic cell and the bacterial ancestor of mitochondria.

Subcontracting of function can lead to the evolutionary diversification of organellar function mediated by changes in nuclear-encoded genes. The evolutionary flexibility of animal mitochondria with respect to modes of energy production is impressive (Tielens et al. 2002). For example, the mitochondria in helminth parasites can switch between aerobic respiration and anaerobic respiration with fumarate as the terminal electron acceptor. The reduction of fumarate is catalyzed by the enzyme fumarate reductase, which has evolved from succinate dehydrogenase, the enzyme that mediates the reverse reaction (the oxidation of succinate to fumarate) in aerobic mitochondria. The mitochondria in certain bivalve mollusks in sulfide-rich anoxic sediments can additionally use inorganic sulfide as electron donor, a trait unknown in conventional aerobic mitochondria which are dependent on organic electron donors. The oxidation of inorganic sulfide to thiosulfate by mitochondria is harnessed to respiratory electron transport and, thereby, both the potentially toxic sulfide is detoxified and ATP is produced.

The evolved condition of a symbiont-derived organelle with functions distinct from those present in its free-living ancestor has an important consequence: that the eukaryotic cell becomes dependent on the organelle for reasons independent of its original function (respiration for mitochondria, oxygenic photosynthesis for plastids). An organelle called the apicoplast found exclusively in the Apicomplexa, a group of parasitic protists that includes the malaria parasite *Plasmodium*, illustrates this point. The apicoplast is bounded by multiple membranes (as expected of organelles with a symbiotic ancestry) and contains a 35 kb circular genome. The sequence of the apicoplast genes provides strong evidence that this organelle is affiliated to plastids, even though it lacks any photosynthesis-related genes (Fast et al. 2001). There is increasing evidence that these nonphotosynthetic plastids are retained because they are the site of fatty acid and isoprenoid synthesis in apicomplexans (Walter and McFadden 2005).

3.6.4 Asset Stripping of Symbiont Function

The protein products of some symbiont-derived genes transferred to the nucleus are not targeted back to the symbiont-derived organelle but are localized to another cellular compartment (figure 3-7d). This process can be described as genetic asset stripping by the host; the host has removed functions from its symbionts to its own advantage.

The extent of asset stripping has been addressed by Gabaldon and Huynen (2003), who identified some 630 nuclear genes in yeast and humans with clear sequence similarity to α-proteobacteria, many (possibly all) of which are derived from the ancestor of mitochondria. More than half of these genes code for cytoplasmic, and not mitochondrial, proteins, including those involved in fructose/mannose metabolism and in lipid and nucleotide synthesis. Similarly, there is strong phylogenetic evidence that genes coding for sucrose synthesis in the cytoplasm of algae and plants (e.g., sucrose phosphate synthetase, sucrose phosphate phosphatase) have evolved from the sucrose biosynthesis genes in the cyanobacterial ancestor of plastids (Lunn 2002).

One predicted consequence of genetic asset stripping by the nucleocytoplasm is that the nucleus of eukaryotic lineages that have secondarily lost an organelle may bear organelle-derived genes with function in the nucleocytoplasm. For example, the anaerobic protist *Trichomonas vaginalis* lacks mitochondria but possesses the chaperonin gene *cpn-60* with close sequence similarity to the mitochondrial *cpn-60* in mitochondriate eukaryotes (Horner et al. 1996).

3.6.5 *Why Is Genetic Assimilation of Symbionts Rare—or Is It?*

This section addresses the exceptional observation that among the large numbers of vertically transmitted symbionts, many with decaying genomes, there are just two definitive instances of genetic assimilation involving the transfer of functional symbiont genes to the host nucleus: the α-proteobacterial ancestor of mitochondria and the cyanobacterial ancestor of plastids.

The obvious explanation for this observation is that genetic assimilation is rare, with the implication that it is evolutionarily difficult. To pursue this explanation further, we need to dissect functional gene transfer into two sequential steps. The first step is the transfer of symbiont DNA to the nucleus. This is "not an improbable event" (Doolittle 1998; see also Timmins et al. 2004) because eukaryotic nuclei are predisposed to take up naked DNA and symbiont cells are occasionally lysed, releasing their DNA into the cytoplasm. However, the transferred DNA persists in the host lineage only if the recipient nucleus contributes to the next host generation. This condition is met in most protists, where host cells (i.e., cells bearing the symbionts) give rise to the reproductive cells, but not in complex multicellular hosts, such as plants and animals, where the host cells are generally somatic cells (e.g., the mycetocytes of insects) destined to die. In multicellular hosts with division between the somatic and reproductive cells, the opportunity for DNA transfer is restricted to brief periods when the symbionts are located in oocytes. Gene transfer is predicted to be much more likely in protists than in multicellular hosts, but the latter is not impossible as is illustrated by the evidence for DNA transfer from the vertically transmitted *Wolbachia* to the nuclear genome in both insects and nematodes (Hotopp et al. 2007; Nikoh et al. 2008).

The second step in genetic assimilation is the accurate transcription and translation of nuclear copies of the symbiont gene and targeting of the protein back to the symbiont. Protein targeting requires the evolution of both a targeting sequence on the protein and import machinery on the symbiont/organelle membranes, and it has been argued that this is the major evolutionary hurdle to the transition of a microorganism to an organelle (Cavalier-Smith 1999). This explanation requires that the cellular controls over the fate of newly synthesized proteins were more fluid in early eukaryotic hosts than in modern eukaryotes, with the expectation that organelles cannot evolve today because the targeting ambiguities and errors inherent in any transitional steps toward the evolution of a novel organelle would be catastrophic to the function of a modern cell. Contrary to this scenario, the targeting sequences in modern cells are not highly specific (Bryan-McKay and Greeta 2007). Multiple sequences can target proteins to one compartment, single-point mutations can cause a change in the subcellular localization of a protein, and some targeting sequences direct proteins to more than one compartment. A further indication of the flexibility of protein targeting in modern organisms comes from research on isolated plastids acquired in a viable condition by ciliates and sea slug mollusks feeding on certain algae. The plastids can persist in the new host for days to weeks, contributing photosynthetic carbon to their hosts. Nevertheless, they gradually decay in many cases because they lack a supply of newly synthesized protein coded by the algal nucleus. In some relationships, however, there is evidence that plastid function and longevity are prolonged by the acquisition of proteins newly synthesized in the host cytoplasm and targeted to the plastids. For the ciliate *Myrionecta rubra* (=*Mesodinium rubrum*), this has been attributed to the retention of both nuclei and plastids of the cryptophyte algal prey; the nuclei are functional and code for plastid-targeted genes that are expressed and then processed in the host cytoplasm (Johnson et al. 2007). In the relationship between the sea slug *Elysia chlorotica* and plastids from the chromophyte alga *Vaucheria litorea*, there is persuasive evidence that at least one plastid protein, PsbO, is coded by the host nucleus (perhaps recently transferred to the host genome), indicating some capacity for the evolution of protein targeting in novel nucleocytoplasm/organelle combinations in modern cells (Rumpho et al. 2008).

The lack of an entirely satisfactory explanation for the apparent rarity of symbiont-derived organelles leads us to consider alternative possibilities.

One is that genetic assimilation is widespread but unrecognized, especially (but not exclusively) among the great diversity of unstudied protists. Marin et al. (2005) have obtained evidence that the photosynthetic structures, known as cyanelles, in the amoeboid protist *Paulinella chromatophora* are of cyanobacterial origin but phylogenetically distinct from plastids. The extent of genetic assimilation of cyanelles by their hosts remains to be established, but these relationships could display some attributes of organelles. Similarly, the genome size of some obligately vertically transmitted bacteria in insects is very small; including *Carsonella ruddii* in psyllids with a genome of 0.16 Mb coding just 182 genes (Nakabachi et al. 2006). Could *Carsonella* have been reduced to an organelle with functions met by bacterial genes transferred to the insect genome? It is not inconceivable that further research will reveal eukaryotes with organelles derived from methanogenic, nitrogen-fixing, or amino-acid-synthesizing symbionts, or with aerobic respiration or photosynthesis obtained through organelles phylogenetically distinct from the known mitochondria and plastids.

Furthermore, genetic assimilation may have gone to completion in many symbionts additional to the hydrogenosomes described in section 3.6.2, leaving either no trace or genome-free membrane-bound compartments. Over the years, symbiotic origins have been invoked for essentially all of the membrane-bound organelles in eukaryotic cells (Martin 2005). Many of the scenarios are not supported by evidence and they are not currently favored. Nevertheless, we cannot exclude the possibility that the eukaryotic cell has evolved from a community of bacteria, with just the mitochondria and plastids retaining the molecular evidence of their ancestry, linked to their requirement for resident genes for functional electron transport chains.

3.7 Résumé

Symbioses persist. This is despite the expectation that the reciprocal exchange of benefit is costly to the participants with the predicted consequences of conflict and propensity for symbioses to collapse. The evidence that at least some symbioses are regularly exploited by cheaters provides further evidence for conflict in symbioses.

I have argued in this chapter that symbioses persist because conflict is managed and because some symbiotic interactions are cost-free. I will consider the latter issue first. The fact that some services are apparently not costly to provide and so incur no conflict is often neglected. Cost-free interactions occur in relationships between organisms with complementary traits, such as the metabolite cross-feeding in microbial consortia (as illustrated in chapter 1, section 1.4.2), methanogenic bacteria which utilize waste hydrogen from anaerobic protists for energy production, and the symbiotic algae which use excretory nitrogen of their animal hosts as a source of nitrogen (section 3.2.3). Although cost-free interactions have not been studied extensively, they may contribute to the persistence of symbioses under circumstances where conflicts over costly interactions are especially intense.

The management of conflict in persistent associations can be attributed partly to transmission mode and its consequences. The more an organism gains from the fitness of its partner, the more beneficial the organism will be to its partner (Ewald 1994; see also chapter 2, section

2.2.3), with obligate vertical transmission promoting the strongest overlap of selective interest between the partners. There is excellent experimental evidence that enforced vertical transmission can promote mutualistic traits. Nevertheless, vertical transmission is neither necessary nor sufficient to resolve symbiotic conflict, as is illustrated, respectively, by the many horizontally transmitted symbioses and the existence of *Wolbachia* and other vertically transmitted reproductive parasites. The other key process involved in conflict resolution is the imposition of rewards and sanctions that control the traits of the partner.

The partner most likely to invest in rewards and sanctions is the one with the least to gain from cheating or the most to lose from a partner that cheats, and this is commonly the host partner. In other words, many symbioses are not associations of equals, but involve one organism that can control many of the traits of its partners. These processes are likely to be particularly important for horizontally transmitted symbioses. Over the years, various authors have queried how horizontally transmitted symbioses persist in apparent harmony without the selective overlap imposed by vertical transmission (e.g., Wilkinson and Sherratt 1999); the solution is sanctions and rewards imposed by the controling partner.

Obligate vertical transmission tends to reduce host-symbiont conflict, but it also promotes conflict among the symbionts over transmission. This conflict can be resolved by host-imposed sanctions and the imposition of tight bottlenecks at transmission. The latter leads to genomic deterioration of the symbionts and further increase in host control over the traits of the symbiont. With genes of symbiont-derived organelles transferred to the host nucleus, host control becomes near-absolute; the host controls whether the products of symbiont-derived organelle are targeted back to the organelle or other cellular compartments (asset stripping) and whether host gene products are allocated to the organelle (subcontracting). There is evidence that the genetic assimilation of the symbiont has gone to completion in one class of organelles, the hydrogenosome, leaving the autonomous membrane-bound organelle as the sole signature of its ancestry as an independent organism. David Smith (1979) has aptly compared the evolution of symbiont-derived organelles to the Cheshire Cat in Lewis Carroll's *Alice in Wonderland.* The Cat "vanished quite slowly, beginning with the end of the tail, and ending with the grin, which remained some time after the rest of it had gone."

CHAPTER 4

Choosing and Chosen: Establishment and Persistence of Symbioses

MANY SYMBIOSES are the most exclusive of clubs. Although symbioses are prevalent in most ecosystems, most organisms do not participate in most symbioses. The root nodules of a pea plant are occupied by one soil-derived bacterium, *Rhizobium leguminosarum* bivar *viciae*, and none of the myriad other soil bacteria. The sediment and water column in which the Hawaiian squid *Euprymna scolopes* lives support a great diversity of bacteria, just one of which, the luminescent *Vibrio fischeri*, colonizes the squid light organ. Even the human gastrointestinal tract which bears more than 1000 microbial taxa is highly selective, such that most microorganisms which enter the gut with food do not persist in the gut habitat.

Why are symbiotic organisms selective in their choice of partners? The core reason is that symbiotic organisms, like all other organisms, live in an antagonistic world containing predators, competitors, and parasites. The establishment and persistence of a symbiosis depend on the capacity of the participating organisms to discriminate suitable partners from the many incompatible, ineffective, cheating, or otherwise deleterious organisms. By choosing well, a symbiotic organism reduces the level of conflict with its partners. In other words, partner choice, described in this chapter, can contribute to minimizing conflict in symbiosis. In chapter 3 (section 3.4), I have already considered other processes, including controls over transmission patterns and rewards/sanctions, that tend to reduce conflict between symbiotic partners.

The selectivity of symbioses raises a second question: who chooses? The partner that chooses most stringently is the one with the greatest selective interest in the choice and the most to lose from associating with an ineffective or deleterious partner. In symbioses between a single large host and many small symbionts, the host is the most likely to make the choice. This is because the symbionts, being small, tend to have high intrinsic rates of increase and so can afford an error in partner choice more readily than the slowly reproducing host. [In the same way, the host is expected to reward or sanction its symbionts rather than the reverse, as considered in chapter 3 (section 3.4.2).]

The purpose of this chapter is to explore the mechanisms by which organisms enter into and maintain symbioses. This chapter is all about choice—how organisms make particular symbiotic choices (section 4.2), why some symbiotic organisms are more selective in their partner choice than others (sections 4.3 and 4.4), and the processes by which the chosen persist with their partner in apparent harmony (section 4.5).

4.2 How Choices Are Made

4.2.1 An Ecological Perspective: The Host as a Habitat

Some symbiotic organisms live within the body of their partner. From an ecological perspective, their host partner is their habitat. In other symbioses, the hosts represent an important part of the habitat of their symbiotic partners, e.g., the plants that provide domatia for ant colonies. As with all other habitats, host habitats are colonized only by organisms able to tolerate the conditions and to consume resources available in that habitat. For example, most regions of the human gut are anoxic, while the human skin surface is oxic, and these conditions exclude obligate aerobes from the gut and obligate anaerobes from the skin surface (Wilson 2005). Unlike other habitats, however, hosts are additionally under selection pressure to encourage colonization by organisms that promote host fitness and to discriminate against colonists that reduce their fitness. These capabilities have variously been called barriers, filters, or locks. Furthermore, host traits can select for a change in the traits of potential colonists, and the resultant change in the traits of the colonists select for a further change in host traits. This is coevolution. The analogy of the host lock is particularly appropriate in this context. Both partners in a mutually beneficial relationship are under selection to achieve the best fit between the host lock and the symbiont key; but the coevolutionary relationship between a host and parasite can be described as the repeated changing of the host locks and testing of new parasite keys.

To illustrate the role of coevolution in shaping the specificity of symbioses, I will consider two relationships between plants and their protective ant colonies: in the *Leonardoxa africana* species complex of understory trees in West Africa and in *Acacia* trees of Central America.

Leonardoxa africana provides domatia for protective ants in swollen twigs. The ants gain access to a domatium by chewing out an entrance hole, called a prostoma, at an unlignified spot on the twig. The size and shape of the prostomata vary among the three members of the *L. africana* complex (Brouat et al. 2001). The prostomata in *L. a. africana* are elongate and very similar in size and shape to the dimensions of

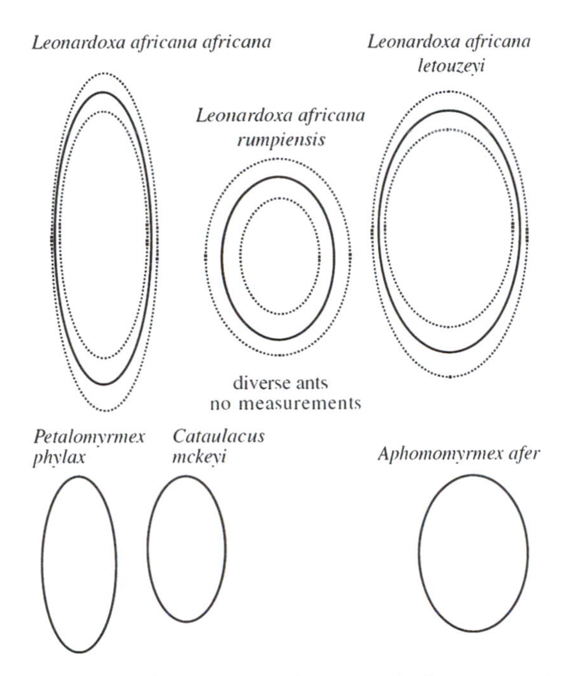

Figure 4-1 Dimensions of prostomata (entrance holes to nest sites for ants) in hollow twigs of *Leonardoxa africana* trees. Mean values are shown as continuous lines and extremes as dotted lines; and the mean dimensions of specialist ant symbionts are displayed [Reproduced from figure 3a of Brouat et al. (2001) with permission from the Royal Society]

the head of the protective ant species *Petalomyrmex phylax*, while the more rounded prostomata in *L. a. letouzeyi* match the dimensions of its ant partner, *Aphomomyrmex afer* (figure 4-1). (The prostomata of *L. a. rumpiensis*, utilized by various ant species, are variable in size and shape.) Brouat et al. (2001) interpret the excellent fit between the prostoma "lock" and ant head "key" in the *Leonardoxa*-ant relationship as coevolutionary morphological adaptations by both the plants and their protective ants to exclude ineffective ant species. However, for *L. a. africana*, this is not entirely successful because an ant cheater, *Cataulacus mckeyi*, can colonize the domatia, as is discussed in chapter 3 (section 3.3).

Central American trees of the genus *Acacia* produce extrafloral nectar (EFN) which is consumed by protective ants, but the composition of the nectar differs between two groups of *Acacia* species: the nonmyrmecophyte species, which produce EFN only when attacked by herbivores and attract opportunistic ants species; and myrmecophyte species, which produce EFN constitutively for their resident ant symbionts (Heil et al. 2005). The EFN of nonmyrmecophytes contains substantial amounts of sucrose, but that of myrmecophytes comprises mostly the monosaccharide sugars glucose and fructose produced by invertase-mediated hydrolysis of sucrose in the nectar (figure 4-2a). The EFN composition of the myrmecophytes is matched by the unusual trait of the resident ant species which, unlike the ants tending nonmyrmecophyte plants, strongly prefer glucose and fructose over sucrose (figure 4-2b) and have very restricted capacity to digest sucrose. In this remarkable relationship, monosaccharide production by myrmecophyte plants and sugar utilization by their ant partners have coevolved in response to the shifting sugar utilization/production traits of the partner. Heil et al. (2005) describe the nectar sugar composition of myrmecophyte species as a chemical filter that discriminates against all but the coadapted ant species.

As these examples illustrate, the framework of host filters or barriers provides a powerful explanation for the specificity of some symbioses. This framework is not, however, sufficient for many associations. There is increasing evidence that some symbioses are underpinned by the exchange of multiple, arbitrary signals, i.e., the signals are functionally unrelated to the information they convey. I consider the role of signals in symbiosis formation next.

4.2.2 *Partner Choice by Signals and Cues*

Signaling in symbioses is often complex, and our understanding of this subject is further complicated by uncertainty and disagreement about the meaning of the term signal. As an approach to minimize semantic difficulties and confusion, I will use the definition of Maynard-Smith and Harper (2003). Although this definition was developed to describe behavioral signaling in animals, it is applicable to symbioses with little change in the wording. A signal is defined by Maynard-Smith and Harper (2003) as an act or structure, which for our purposes includes a molecule, with three characteristics:

1. it alters an attribute of other organisms (an attribute may be the pattern of gene expression, metabolism, behavior, etc.);
2. it has evolved because of its effect on the other organisms; and
3. it is effective only because the response of the recipient has also evolved.

Signals are expected to play a central role in the formation of symbioses, in a similar fashion to their role in the development of multicellular organisms or animal courtship. This is because of two key features of signals. First, they enable organisms (or cells) to transmit precise information about their condition or status. As a result, they can orchestrate

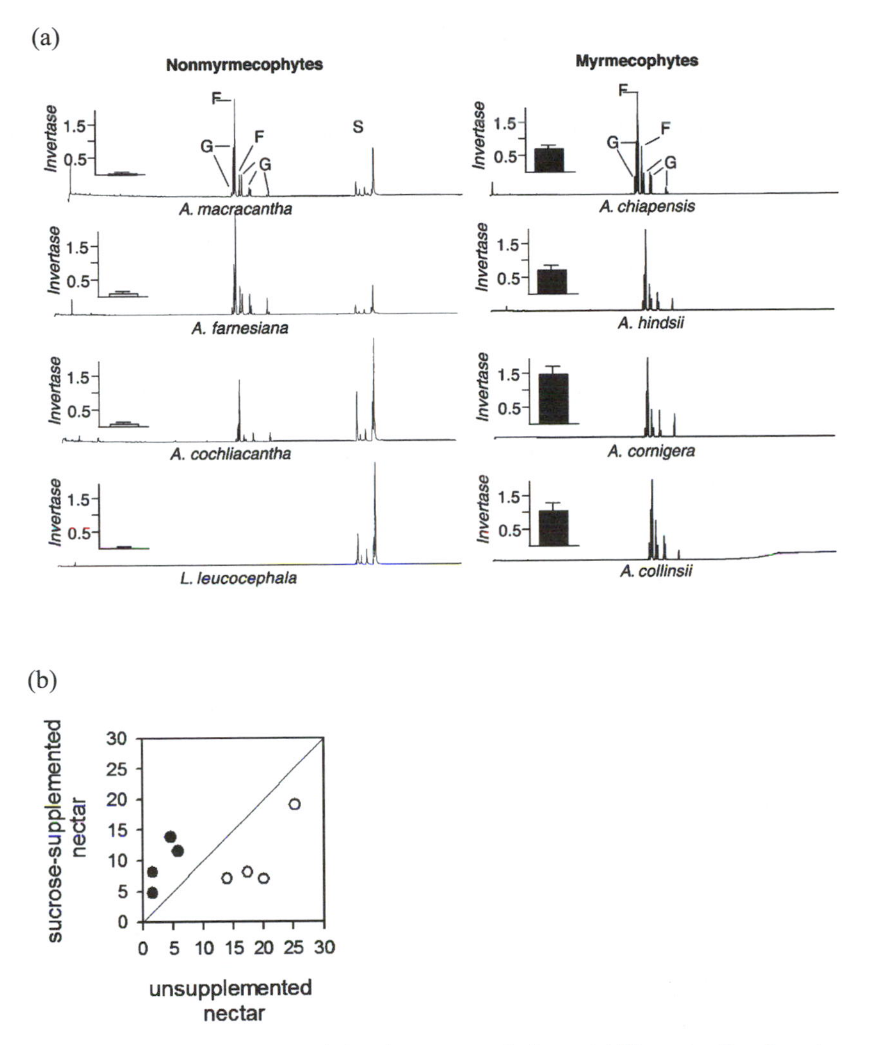

Figure 4-2 Sugar relations of *Acacia*-ant associations. (a) The extrafloral nectar of nonmyrmecophyte (left) and mymecophyte (right) species of *Acacia*. The sugar profiles include peaks for the monosaccharides glucose (G) and fructose (F) and the disaccharide sucrose (S), and 1 unit of invertase activity is 1 µg glucose μl^{-1} nectar min^{-1}. (b) Percentage of ants responding in a choice test to the extrafloral nectar of myrmecophyte *Acacia* species, either supplemented with sucrose or without any sucrose supplement. The eight ant species tested either associate opportunistically with nonmyrmecophyte plants (closed symbols) or associate persistently with myrmecophyte plants (open symbols). [Reproduced from *Science*, figure 2 and redrawn from data in figure 3 of Heil et al. (2005). Reprinted with permission from AAAS]

complex patterns of change in strictly defined spatial and temporal order. Second, signal exchange can occur only between consenting partners. Although the information conveyed may not reflect the condition of the signaler perfectly, the receiving organism cannot repeatedly or generally be deceived or coerced by a signal.

In the symbiosis literature, the term "signal" is sometimes used interchangeably with the term "cue." This is misleading because cues have not evolved as adaptations for symbiosis. An example of a cue is ammonia released as an excretory product by alga-free juveniles of the jellyfish *Cassiopeia* spp. Motile, free-living cells of the dinoflagellate alga *Symbiodinium* swim actively up the ammonia concentration gradient, so promoting contact with potential hosts. Similarly, some bacteria inadvertently release fragments of cell wall peptidoglycan as they grow and divide; the production by luminescent *Vibrio* bacteria of one particular fragment, a disaccharide-tetrapeptide monomer of peptidoglycan commonly known as tracheal cytotoxin, is used by the squid host of the *Vibrio* as a cue to initiate the formation of the light organ that ultimately houses the luminescent bacteria (Koropatnik et al. 2004). Neither the ammonia nor the tracheal cytotoxin is a signal.

Research on signal exchange underpinning symbiosis formation has been studied intensively for associations between legumes and nitrogen-fixing rhizobia. This research has yielded two conclusions of general significance: that signal exchange is intrinsically complex, meaning that the formation of some symbioses cannot be explained in terms of single traits of a partner (section 4.2.3); and that certain signaling modules have been recruited by multiple symbioses (section 4.2.4). For general reviews of symbiosis formation in legumes, the reader is referred to Gage (2004) and Jones et al. (2007).

4.2.3 Signaling in the Legume-Rhizobium Symbiosis

The interactions between many compatible legume-rhizobia combinations are initiated by a three-component exchange of signals, **flavonoid-Nod–factor-kinase** from plant to rhizobium to plant. The flavonoid signal is exuded constitutively from plant roots. A compatible rhizobium responds by synthesizing a Nod factor (an acylated oligosaccharide) which binds to a plant receptor on the cell membrane of the root hair. The receptor is a serine/threonine receptor kinase with one or more LysM motifs predicted to bind the acetyl-glucosamine backbone of the Nod factor. The three-step signal exchange triggers a signaling cascade in the plant cell (figure 4-3), leading to changes in plant gene expression that orchestrate morphological changes required for symbiosis formation.

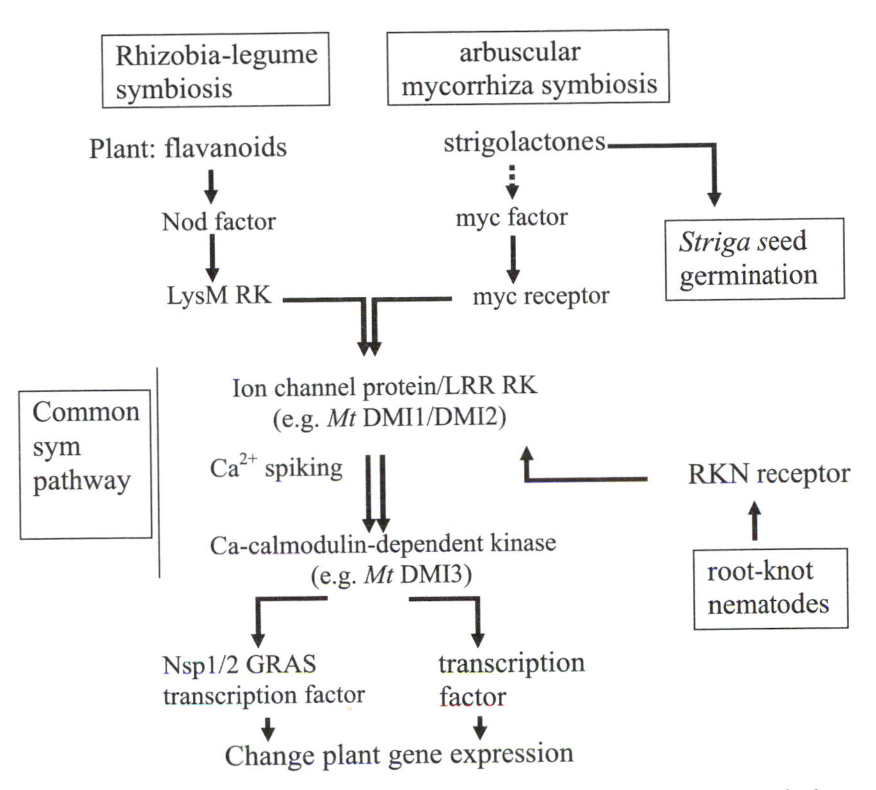

Figure 4-3 Signaling pathways shared by symbionts and parasites of plant roots (see text at section 4.2.3 and 4.2.4 for details) (LRR RK: leucine-rich repeat receptor kinase; *Mt* DMI1-3 are genes of the common sym pathway in *Medicago truncatula*; orthologs of these genes have been identified in other legume species (e.g., NORK genes in *M. sativa*, SYMRK and SYM genes in *Lotus japonicus*, SYM genes in *Pisum sativa*)

Importantly, the flavonoid-Nod–factor-kinase signal exchange is neither universal nor sufficient for symbiosis formation. Two strains of rhizobia are now known that lack the *nod* genes required for Nod factor synthesis, and they can nodulate their native hosts probably by a cytokinin-like signal, although the detail remains to be established (Giraud et al. 2007). Cytokinin signaling may also contribute to symbiosis formation in symbioses dependent on Nod factors, and we can be reasonably confident that many, perhaps all, of these symbioses are underpinned by other molecular interactions and signaling pathways that occur in parallel or interacting with the pathway in figure 4-3. One group of plant molecules implicated in signaling at initial contact are lectins which bind to carbohydrate moieties on the cell wall surface of compatible rhizobia. Experiments with transgenic plants expressing heterologous lectins reveal that the lectin can be a very important determinant of symbiosis formation. For example, alfalfa plants do not normally form a symbiosis with *Rhizobium leguminosarum* bv. *viciae*, the symbiont of pea plants, but this heterologous symbiosis can be generated if, first, the alfalfa plants are modified to bear the pea lectin and, second, *R. leguminosarum* bv. *viciae* is modified to synthesize the Nod factor of the natural alfalfa symbiont, *Sinorhizobium meliloti* (van Rhijn et al. 2001).

The rhizobia also make multiple contributions to signal exchange. As well as producing a Nod factor, they inject various proteins into host cells via their type III secretion system (T3SS). (T3SSs are essentially molecular syringes by which effector proteins are transported from bacteria into eukaryotic cells. I consider their role in symbioses further in section 4.5.3.) Curiously, genetic abolition of T3SS function has variable effects on the capacity of rhizobia to form a symbiosis. Marie et al. (2001) describe the capacity of wild-type and nonfunctional T3SS mutants of the strain NGR234 to nodulate various plants. The mutation had no effect on infection of *Lotus japonicus*, abolished nodule formation in *Vigna unguiculata*, and yielded normal, nitrogen-fixing nodules on *Crotalaria juncea*, in which the wild type induces nonfixing nodules. These results suggest that the proteins secreted via the T3SS render the symbiosis in *C. juncea* incompatible (one possibility is that they induce plant defense responses) but are required for the symbiosis in *V. unguiculata* (they may signal that the rhizobium is acceptable or suppress plant defenses). Several of the proteins secreted by the rhizobial T3SS have been identified, but their modes of action remain to be established.

Taken together, these studies indicate that the processes underlying symbiosis formation in the legume-rhizobium association are both complex and variable among species. Why so complex and variable? Perhaps the organisms are using multiple criteria to assess the quality of potential partners, and these criteria vary among species because they are subject to selection in response to such factors as partner availability and the incidence and traits of cheaters or other deleterious organisms. Although models of symbiosis formation based on single lock-and-key or code-and-password concepts provide adequate explanations for some associations (see section 4.2.1), they are inadequate for interactions between legumes and rhizobia. Other symbioses have not generally been subjected to such intense research, but it is likely that symbiosis formation and partner choice in very many symbioses are underpinned by multiple, complex, and variable signaling events.

In one important respect, however, symbiosis formation appears to be remarkably uniform, not only among the legume-rhizobium symbioses but also across other associations involving plant roots. There is now strong evidence that the plant signaling networks which orchestrate the formation of different associations include a common, ancient module; this module has been captured from a preexisting function and then used, with some modification or remodeling, in novel functions. I consider this ancient signaling module next.

4.2.4 *Signaling and the Establishment of Plant Root Symbioses*

The common sym pathway shown in figure 4-3 is the focus of research on signaling modules shared across different symbiotic systems in plant roots. Genetic dissection of different symbioses has shown that this signaling module, which was first identified in legume-rhizobia symbioses, is required for plant symbiosis with arbuscular mycorrhizal fungi (AM fungi) and the nitrogen-fixing actinomycete symbiont *Frankia* (Gherbi et al. 2008; Stracke et al. 2008). Since the mycorrhizal association is far more ancient than the other symbioses, the sym pathway is presumed to have evolved in relation to the AM symbiosis and subsequently been captured by the other symbioses.

There is an expectation that the AM symbiosis is initiated by plant-to-microbe-to-plant signal exchange analogous to the flavonoid-Nod–factor-kinase system in legume-rhizobial symbioses. To date, the plant signal for AM fungi has been characterized—it is a strigolactone (Akiyama et al. 2005)—but the responding "myc factor" of the fungus and the plant receptor remain to be identified. Furthermore, parts of the signaling pathways mediating the formation of arbuscular mycorrhizas have also been exploited by parasites (figure 4-3). The strigolactones in plant root exudates are detected by the plant parasite *Striga*, and infection of plant roots by root knot nematodes is mediated by the common sym pathway (Weerasinghe et al. 2005).

The common sym pathway poses a problem. How do infections mediated by the same pathway have different outcomes for different partners? For example, how do nodules develop on plant roots when the sym pathway is triggered by the Nod factor but not by the myc factor? We cannot answer this question definitively, but the mode of signal transduction from the proteins DMI1/DMI2 to DMI3 in the common sym pathway of *Medicago* (and their homologs in other plants) may be crucial. DMI3 is a calcium-calmodulin kinase, a protein which undergoes conformational change in response to changes in cytoplasmic Ca^{2+} levels. The cytoplasmic Ca^{2+} levels oscillate in response to infection by both compatible rhizobia and AM fungi, but the pattern of oscillations depends on the symbiont: it comprises regular spiking about once per 100 seconds for the rhizobia, and it is less regular and with higher frequency for the AM fungi (Kosuta et al. 2008). The DMI3 protein is believed to decode the different patterns of Ca^{2+} oscillations in different ways, so triggering different patterns of gene expression in plant cells exposed to rhizobia and AM fungi.

In summary, an ancient symbiotic signaling module known as the common sym pathway is believed to have evolved in the relationship between plants and AM fungi, and to have been exploited by variously more recently evolved symbionts and parasites (Markmann et al. 2008).

The common sym pathway of plants is not only of academic interest. It also creates the biotechnological opportunity to construct symbioses between nonlegumes and nitrogen-fixing rhizobia and, thereby, to reduce the need for chemical nitrogen fertilizer for such major crops as wheat and rice (Dey and Datta 2002; Ladha and Reddy 2003). Rhizobia can invade nonhosts (Perrine-Walker et al. 2007), and root swellings occupied by bacteria can be induced in nonlegumes exposed to rhizobia, for example by exposure to cell-wall-degrading enzymes or a strong magnetic field. However, the cellular architecture of these swellings is disorganized and the nitrogen fixation rates are very low (de Bruihn et al. 1995). The realization that elements of the plant signaling pathway responsive to rhizobia are very ancient, apparently present in all plants capable of forming a mycorrhizal symbiosis, raises the possibility of engineering the full signaling pathway for rhizobial symbiosis in nonlegume crops. In principle, such engineered plants would support the infection and persistence of an effective nitrogen-fixing symbiosis.

4.3 How Specific Are Symbioses?

Specificity refers to the taxonomic range of partners with which an organism associates. Any assessment of the specificity of an organism for its partner(s) must be qualified by the method adopted to determine specificity. There are essentially two approaches: field, also known as ecological, and experimental, also known as physiological. The field approach determines the diversity of taxa with which the organisms of interest associate in the natural environment, and the experimental approach challenges the organisms with potential partners under defined conditions. The two methods can yield very different conclusions about specificity, as is illustrated by studies of arbuscular mycorrhizas. The AM fungi were traditionally believed to be exceptionally promiscuous—any AM fungus can infect all plants capable of forming

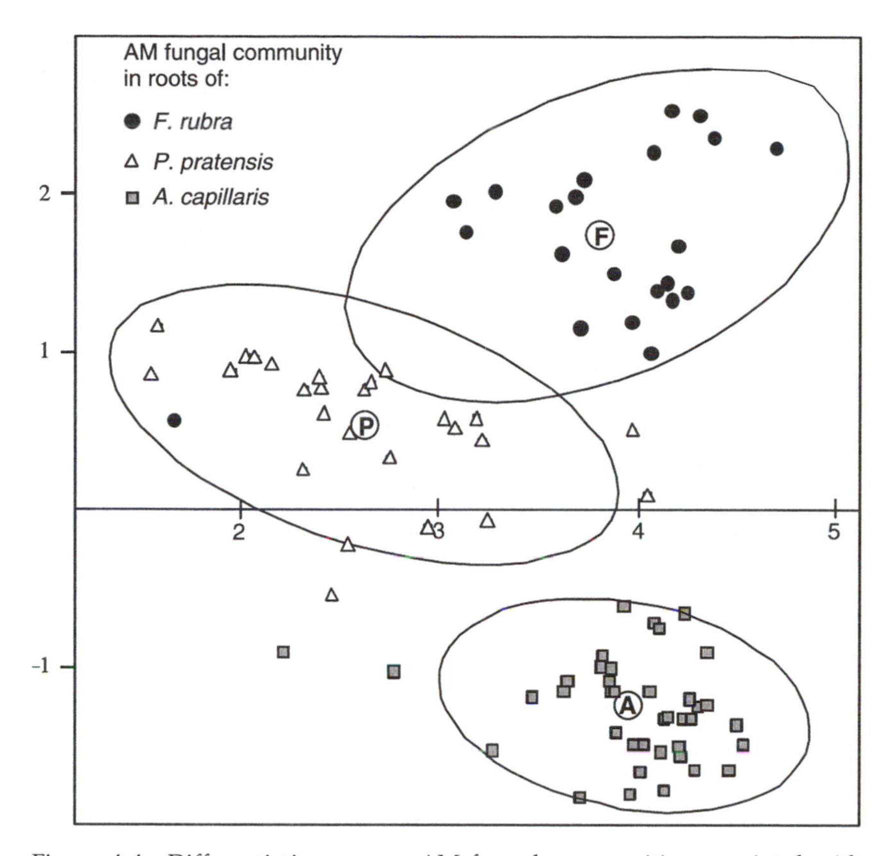

Figure 4-4 Differentiation among AM fungal communities associated with three co-occurring grass species, *Agrostis capillaris* (A), *Festuca rubra* (F), and *Poa pratensis* (P), as revealed across two axes of principal components analysis. Each point represents the AM fungal community in a single root, characterized by the profile of terminal-restriction fragment lengths. For each host plant species, confidence ellipses ($P < 0.05$) are shown, and the labeled circles indicate the center of gravity (mean value). [Reproduced from figure 2 of Vandenkoornhuyse et al. (2001) with permission from Wiley-Blackwell]

arbuscular mycorrhizas (Smith and Read 2007). This viewpoint arose principally from pot experiments with individual plants grown in pots and infected with fungal isolates from laboratory spore cultures. It has not been confirmed by studies of AM fungi in natural vegetation, which reveal a very different relationship, such that different plant species in a single habitat tend to be infected by different communities of fungi. For example, three grass species co-occurring on a single hill pasture plot in Scotland are associated with statistically different communities of AM fungi (figure 4-4).

Various factors might contribute to the different conclusions from laboratory experiments and field surveys. The experimental approach tends to focus on taxa which can be maintained readily in the laboratory (for AM fungi, the prolific spore-producers). These are not necessarily widespread, or even present, in the field condition, and they may have very different specificity profiles from the dominant naturally occurring taxa. A particularly vivid cautionary tale comes from the realization that most of the cultured *Symbiodinium* strains used in laboratory trials of specificity in corals and related marine animals are contaminants, while the dominant symbionts in many hosts are unculturable (Santos et al. 2001).

Field analyses can also be difficult to interpret. They tend to underestimate the range of compatible partners because only a limited diversity of partners is available to the host, and they can yield misleading results as a consequence of inadequate sampling. The problem of inadequate sampling can be illustrated by one of the most intensively studied field analyses of specificity: on the microbiota in the human gastrointestinal tract. As well as being diverse (>1000 bacterial taxa), this microbiota appears to be remarkably specific to each person, "as unique as a fingerprint" (Dethlefsen et al. 2007). For example, in one analysis of the microbiota associated with the intestinal wall of 192 patients, nearly all of 80 bacterial sequences from each person were unique to that individual (Frank et al. 2007). The most straightforward interpretation of these data is that the associations are exquisitely specific, perhaps as a result of the unique nutritional physiology, immunology, etc. of each human (additionally, some microbial colonists might evolve in situ under the unique selection pressure exerted by each human genotype). However, inadequate sampling could account for this apparent specificity, in two ways. Perhaps many microbial taxa are very widely distributed (i.e., the symbiosis is relatively nonspecific), but this would not be detected if their relative density varied among individual hosts and only the most abundant taxa were detected. Alternatively, the interhuman variation may be real, but caused by chance and not specificity. As Curtis and Sloan (2004) have argued, an individual host can be considered to sample the microbial community in its environment. If the microbial diversity in the environmental community is very large, then a single host is most unlikely to acquire all the taxa with which it is compatible. It may take many thousands of host individuals for all representatives of the microbial metacommunity to be recorded.

These various caveats notwithstanding, there is good evidence that symbiotic species vary in their specificity in the field. Quantitative data are available for the ectomycorrhizal fungi (ECM fungi) which produce

TABLE 4-1
Plant Range of Ectomycorrhizal Fungi

(a) Plant Range of Fungal Species of Different Families

Family of fungus	*Number of fungal species*		
	Narrow plant range	*Intermediate plant range*	*Broad plant range*
Amanitaceae	10	70	120
Boletaceae	156	142	67
Cortinariaceae	365	743	356
Gomphidiacaeae	20	11	0
Hygrophoraceae	88	125	37
Russulaceae	257	345	200
Sclerodermataceae	8	15	8
Tricholomatacaeae	35	68	76
Total	939 (28%)	1519 (46%)	864 (26%)

(b) Plant Range of Fungal Species with Different Modes of Spore Dispersal

Mode of dispersal	*Number of fungal species*		
	Narrow plant range	*Intermediate plant range*	*Broad plant range*
Above ground	743 (25%)	1342 (46%)	864 (29%)
Below ground	196 (53%)	177 (47%)	0

The values are calculated from data in Molina et al. (1992). Narrow plant range: restricted to a single plant genus; intermediate plant range: colonizes members of one to several host families (but not both angiosperms and gymnoperms); broad plant range: associated with a diversity of plants, including both angiosperms and gymnosperms.

conspicuous and readily identifiable toadstools or truffles. Using data on the plant range of more than 3000 ECM fungal species, Molina et al. (1992) found that just over a quarter of the species have a very broad plant range, associated with both angiosperms and gymnosperms, a similar proportion is restricted to a single plant genus, and the specificity of the remainder is intermediate (table 4-1a). Examples of extremes of plant range include *Cenococcum geophilum*, associated with most plant genera that can form ectomycorrhizas, and *Suillus cavipes*, described only with *Larix*. Limited data indicate that specificity also varies widely among different host species in some other symbioses, e.g., *Symbiodinium* in corals, rhizobia in legumes (LaJeunesse 2002; Gage 2004).

From an ecological perspective, one possible explanation for the observed variation in specificity is a trade-off between the risk of failing to find a partner through high specificity, and the risk of associating with an ineffective or deleterious partner through low specificity. In principle, one would expect high specificity to be affordable if the organism can persist in isolation or if the preferred partner is readily available. However, it can be difficult to quantify partner availability, which is determined by the probability of contact and may not be correlated well with the partner's abundance in the environment. Two examples illustrate how specificity and partner availability can be linked. The first is the pollination mutualism of the creosote bush *Larrea tridentate*, which lives in the arid deserts of North America and flowers in response to irregular rainfall. At first sight, creosote bush flowers appear to be a most unpredictable resource for pollinating insects, suitable only for generalist floral visitors. In fact, the flowers of *L. tridentata* are visited by up to 120 species of solitary bees, including many specialists which utilize no other plant species. The bees emerge from diapause in response to rainfall, and so in synchrony with flower production (Danforth 1999). In other words, the pollinating bees can afford to be highly specific because the availability of their partner can be predicted by rainfall. The second example is lichenized fungi that reproduce by sexual spores. On germination, a spore develops into a lichen only if it contacts a suitable alga or cyanobacterium, and these symbionts are generally very rare in the free-living condition, presumably because the small number of viable cells released from lichens do not persist indefinitely in isolation. Rikkinen et al. (2002) have proposed that this problem of symbiont supply is ameliorated for communities of lichens with a stable species composition, such as the epiphytic lichens in ancient woodland. All the lichen species in the community are compatible with the same small pool of algal genotypes, and this shared resource promotes the probability that fungal propagules encounter a suitable symbiont. In this system, low symbiont abundance has led to narrow field specificity, albeit a specificity for the most available genotypes. It remains to be established whether these lichenized fungi also have a narrow experimental specificity and whether the monopoly of the few symbionts is stable over long periods.

The organisms in some symbioses display transmission mechanisms, i.e., adaptations that enhance the probability of contact between the partners. Generally, efficient transmission is correlated with narrow specificity. Among ECM fungi, spore dispersal is mediated either abiotically by air currents from above-ground toadstools, puffballs, etc. or by mycophagous animals from below-ground truffles. The probability of transmission to a compatible plant is far higher for the animal-transmitted spores than for those transmitted abiotically; and the ECM fungi with below-ground fruiting bodies tend to be more specific than those with above-ground structures (table 4-1b).

I conclude by considering the specificity of vertically transmitted symbioses. With obligate vertical transmission, the diversity of symbionts available for a host is reduced to the complement in the parent, and this would lower the selection pressure on the host to discriminate among potential symbionts. Does this mean that the experimental specificity of these vertically transmitted symbioses is very broad through relaxed selection, or narrow through coevolution such that organisms are compatible only with the inherited lineage of symbionts? This problem has not been explored extensively, largely because most vertically transmitted symbioses are so intimate that they cannot be manipulated. Exceptionally, the symbiosis between one group of insects, the plataspid stinkbugs, and their vertically transmitted gut bacteria *Ishikawaella* can be manipulated because the female deposits a capsule containing *Ishikawaella* cells next to each laid egg, available to be consumed by newly hatched offspring. When Hosokawa et al. (2007) switched the capsules deposited by females of two species of *Megacopta*, the symbioses in both species developed apparently normally with the heterologous symbiont. In *Megacopta*, the system is not absolutely species-specific, but it is not known whether phylogenetically more distant combinations of plataspid host and symbiont are compatible.

The specificity of vertically transmitted symbioses is not just of academic interest. Enucleated nonhuman eggs, especially rabbit eggs, have been proposed as recipients of human nuclei in somatic cell nuclear transplantation experiments (also known as therapeutic cloning) to produce early embryos and embryonic stem cells (Fulka et al. 2008). In addition to the possible dysfunction of a nucleus introduced to a cytoplasm of some 100,000 heterologous proteins (Gurdon and Byrne 2003), this proposed procedure is underpinned by the untested assumption that the donor nucleus is compatible with the mitochondria of the enucleated recipient cell, despite the divergent evolutionary history of nucleocytoplasmic-mitochondrial interactions between humans and rabbits over millions of years. Circumstantial evidence for specificity in these vertically transmitted systems comes from the observation that rat/mouse and sheep/goat hybrids have a poorly functioning electron transport chain, low ATP production, and limited developmental capability (Fulka et al. 2008).

4.4 Insights from Mixed Infections

4.4.1 Predicted Consequences of Mixed Infections

So far, I have made the simplifying assumption that symbioses involve just two partners, often a host and its population of genetically uniform symbionts. This does not describe adequately the many symbioses with more than two partners. The implications of multiple partners have been explored for parasites infecting a single host, and the conclusions are relevant for symbioses. There are two broad models.

The first model supposes that the different parasite genotypes compete and, consequently, exploit host resources more quickly than if they were in single infections (van Baalen and Sabelis 1995; Frank 1996). This is deleterious to the host if, as is commonly assumed, virulence and competitiveness are correlated. This model is a form of the generic problem of the Tragedy of the Commons: a limiting resource is shared by individuals whose fitness is defined by their own traits and not by the optimal strategy for exploiting the resource. The problem is exemplified by the conflicts inherent to the medieval agricultural practice of communal grazing of livestock or some modern fishing practices in seas with limited fish stocks, where it is in the individual interest to graze or fish at unsustainable levels (Hardin 1968). For single-genotype infections of parasites, those genotypes which exploit the host rapidly are less fit than prudent genotypes which exploit the host slowly, but in mixed infections their rapid consumption of host resources leaves the fitness of the prudent genotypes disproportionately reduced. Extending this model to symbioses, mixed infections are predicted to promote competition, selection for aggressive exploitation of host resources, and reduced benefit to the host.

The second model (Chao et al. 2000) is a form of a different problem known as the Collective Action model. This is illustrated by the familiar conflict inherent to all tax systems: the collective good of group services funded by taxes is counterbalanced by the individual cost of paying those taxes because each individual does not have exclusive access to their own contribution. For many parasites, gene products, such as enzymes or other virulence factors, that collectively promote host exploitation are the equivalent of taxes. Multigenotype infections are predicted to select for self-interest over group interest, resulting in cheaters (also known as free-loaders) that do not display the group trait but benefit from the exploitation by the host. For parasite-host interactions, the Collective Action problem leads to reduced virulence, the reverse of the prediction from the Tragedy of the Commons (above). For beneficial symbioses, however, the predicted consequence of the Collective Action model is the same as for the Tragedy of the Commons: depressed host fitness. This is because many group traits of symbionts are advantageous to the host, such as nutrient provisioning or protection from predators.

In summary, verbal arguments derived from models of virulence evolution in parasites suggest that mixed infections of symbiotic organisms are likely to be deleterious to the host, whether they cause conflicts of the Tragedy of the Commons or Collective Action. To what extent are these models relevant to real symbioses?

The Tragedy of the Commons model is founded on competition among symbiotic organisms for access to common resources. There is excellent evidence for competition in various symbioses, including the ant-plant associations and bacteria in mammalian guts. Ants compete for domatia on myrmecophyte plants. For example, the East African tree *Acacia drepanolobium* studied by Palmer (2004) associates with four ant species (*Crematogaster mimosae*, *C. nigriceps*, *C. sjostedti*, and *Tetraponera penzigi*), but each tree bears a single ant species. When ant colonies on two adjacent trees came into contact by tree canopy overlap, they fight and the usual outcome conforms to a linear dominance hierarchy: *C. sjostedti*>*C. mimosae*>*C. nigriceps*>*T. penzigi*. This hierarchy is not dictated by interspecific variation in the fighting ability of the individual worker of the different species but by sheer force of numbers; a larger colony can better tolerate the mortality. Thus, *C. sjostedti* has larger colonies than the other species and generally wins fights, *C. mimosae* has larger colonies than *C. nigriceps*, and so on (figure 4-5a). When Palmer (2004) experimentally reduced the colony size of the competitive dominant species, the outcome of the subsequent fight was reversed (4-5b).

Competition among *Escherichia coli* strains in the mammalian gut is mediated chemically by proteinaceous toxins known as colicins that kill bacterial strains lacking the capacity to synthesize both the colicin and its antidote (known as the immunity protein). Among *E. coli* strains, some 25 colicins are known, and they variously kill cells by degrading DNA, inhibiting protein synthesis, or destroying cell membrane integrity. When a colicin-producing strain of *E. coli* (C) is co-cultured with a colicin-susceptible strain (S) lacking the capacity to produce the colicin and immunity protein, the C strain invariably outcompetes the S strain, which is eliminated (Riley and Gordon 1999). The outcome is more complex in *E. coli* populations that include strains which can produce the immunity protein but not the colicin; these are called resistant (R) strains. The R strains are competitively subordinate to S (R incurs the cost of producing the immunity protein), but dominant to C (because it does not have the cost of producing the colicin), creating an intransitive dominance network (i.e., no single strain is dominant over all competitors) with C>S, S>R, and R>C. In both modeled and experimental populations, none of the strains is eliminated, and the three types compete indefinitely (Kirkup

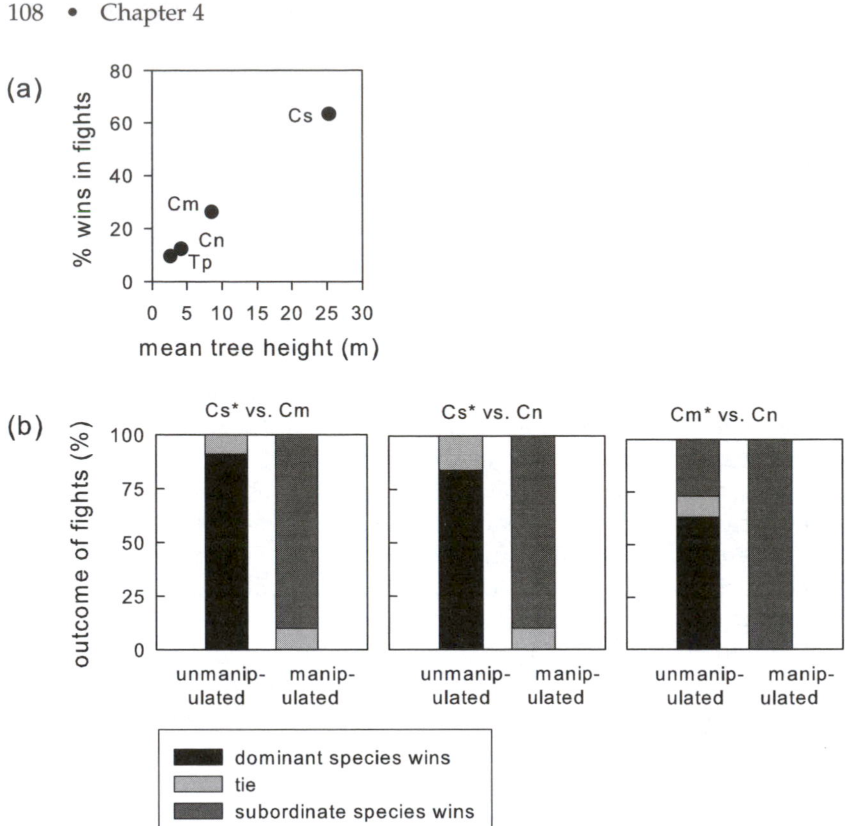

Figure 4-5 Competition by fighting among ant species (Cs: *Crematogaster sjostedti*, Cm: *C. mimosae*, Cn: *C. nigrescens*, Tp: *Tetraponera penzigi*) that colonize the tree *Acacia drepanolobium*. (a) Positive correlation between mean tree height (index of colony size) and % interspecific fights that each species wins. (b) Outcome of staged fights in which colony size of the dominant (*) and subordinate ant species is either unmanipulated or manipulated to reduce the colony size of the dominant species. [Reproduced from *Animal Behaviour*, figure 5 of Palmer (2004) with permission from Elsevier]

and Riley 2004). Competitive interactions among strains of a second gut bacterium, *Bacteroides fragilis*, are also complex. Central to competition in this species are the surface polysaccharides and glycoproteins which generally include the pentose sugar fucose. Cells of a mutant *B. fragilis* lacking the capacity to fucosylate surface molecules are able to form a symbiosis in the absence of the wild-type *B. fragilis*, but they are outcompeted and lost from the gut when the wild-type cells with surface fucose residues are present (Coyne et al. 2005). The competitiveness of the fucosylated *B. fragilis* is, however, dependent on the host. The bacteria acquire fucose by cleaving terminal fucose moieties from glycoproteins and glycolipids in the extracellular matrix of intestinal cells. The important twist to this interaction is that host cells only become fucosylated in the presence of *B. fragilis*, meaning that the wild-type *B. fragilis* induces the host to promote its competitiveness (Coyne et al. 2005).

In other symbioses, there is excellent evidence for a central prediction of the Collection Action model: the existence of cheaters that provide little or no benefit to their partners. The symbioses between animals and photosynthetic algae appear to be particularly susceptible. Most symbiotic algae release a substantial proportion of the photosynthetically fixed carbon to the host, but some genotypes release little carbon, resulting in increases in their proliferation rates and reduced growth and vigor of their hosts. This has been demonstrated for *Chlorella* in laboratory cultures of freshwater hydra (Douglas and Smith 1984), prasinophyte algae in field populations of the intertidal flatworm *Symsagittifera* (*Convoluta*) *roscoffensis* (Douglas 1987), and *Symbiodinium* in various corals (Stat et al. 2008). Other symbiotic organisms that cheat by failing to provide a service to their partners are described in chapter 3 (section 3.3).

These data do not, however, provide a complete validation for either of the models in relation to symbioses. There is a dearth of data that explicitly test crucial predictions of both models: for the Tragedy of the Commons, that the interpartner competition is deleterious for the host in a symbiosis; and for the Collective Action model, that the cheating traits evolve in the context of mixed infections. Furthermore, some data sets suggest that the impact of multiple partners on real symbioses can be more variable than these models predict. An important set of experiments was conducted by van der Heijden et al. (1998) on symbioses between plants and AM fungi. Eleven plant species were infected with four AM fungi, either individually or simultaneously, and eight of them displayed a growth response to the symbiosis. The prediction from the models that the mixed infection would give a lower plant biomass than all individual infections was not obtained for any of the plant species. For one species, the mixed infection treatment supported a greater plant biomass than all the individual infections (the reverse of the prediction), and for the remainder, the impact of the mixed infection was intermediate (figure 4-6). Although the reasons for the among-plant variability were not investigated, these data indicate that the impact of mixed infections can be strongly context dependent, varying with the specific identity of the interacting partners.

An important limitation of the models is that they assume that the multiple taxa in a mixed infection are functionally equivalent, in the sense that the host derives the same type of benefit from the alternative taxa. For many symbioses, the multiple partners are functionally distinct,

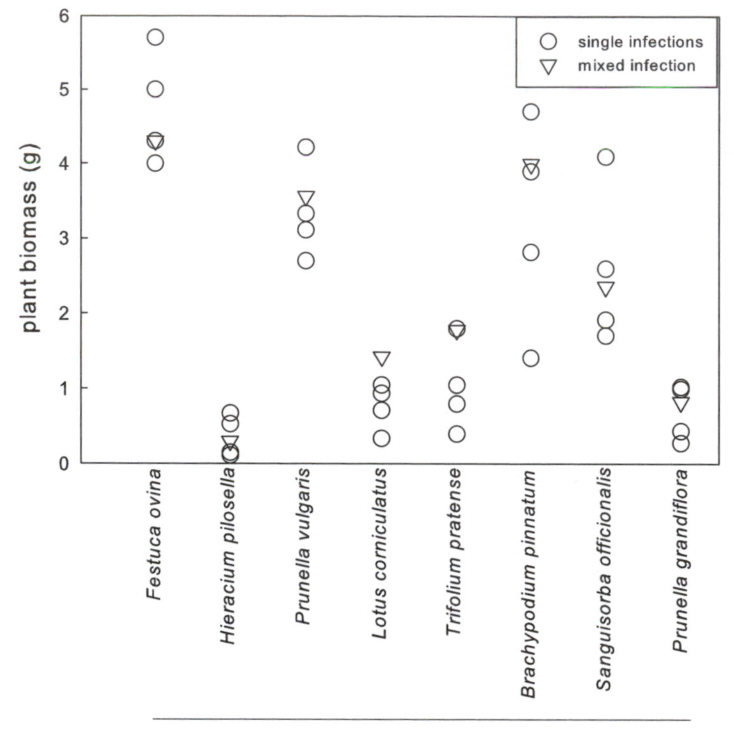

Figure 4-6 Biomass of plants infected with four AM fungi either as single infections or as a mixed infection. [Adapted by permission from Macmillan Publishers Ltd.: Nature, from figure 1 of van der Heijden et al. (1998), copyright 1998]

with very different predicted consequences for a mixed infection. I consider this issue next.

4.4.2 *Mixed Infections with Functionally Distinct Partners*

There is strong evidence that organisms can benefit from accommodating multiple partners with different traits. Consider the intertidal red alga *Chondrus crispus* which, in the field, is always colonized by two snail species, *Anachis lafresnayi* and *Mitrella lunata* (Stachowicz and Whitlach 2005). Field experiments on the alga caged with one, both, or neither of the snail species revealed that the alga was protected from different groups of fouling organisms by the two snail species: from bryozoans by *M. lunata* and from solitary ascidians by *A. lafresnayi*. Because of their complementary protective functions, the alga requires both snail species under field conditions.

Functional complementarity has also been suggested for multiple bacterial symbionts in various animals. For example, genome analysis has predicted complementary metabolic capabilities of two vertically transmitted bacteria, *Baumannia cicadellinicola*, a γ-proteobacterium, and *Sulcia muelleri*, a member of the Bacteroidetes, in an insect, the glassy winged sharpshooter *Homalodisca coagulata*. *Sulcia* can synthesize essential amino acids and *Baumannia* can synthesize various vitamins and cofactors, all nutrients in short supply in the insect's diet of xylem sap (McCutcheon and Moran 2007). Furthermore, the putative capabilities of the two bacteria are interdependent. For example, *Sulcia* produces homoserine, the substrate for synthesis of the essential amino acid methionine by *Baumannia*, and *Baumannia* provides the polyisoprenoids required for menaquinone synthesis by *Sulcia* (figure 4-7). In other words, there is cross-feeding of metabolites between the two bacteria, and the insect host is in symbiosis with a consortium of two bacteria with complementary functions. (As explained in chapter 1, section 1.4.2, bacteria with complementary metabolic capabilities that display cross-feeding of metabolites to their mutual benefit are

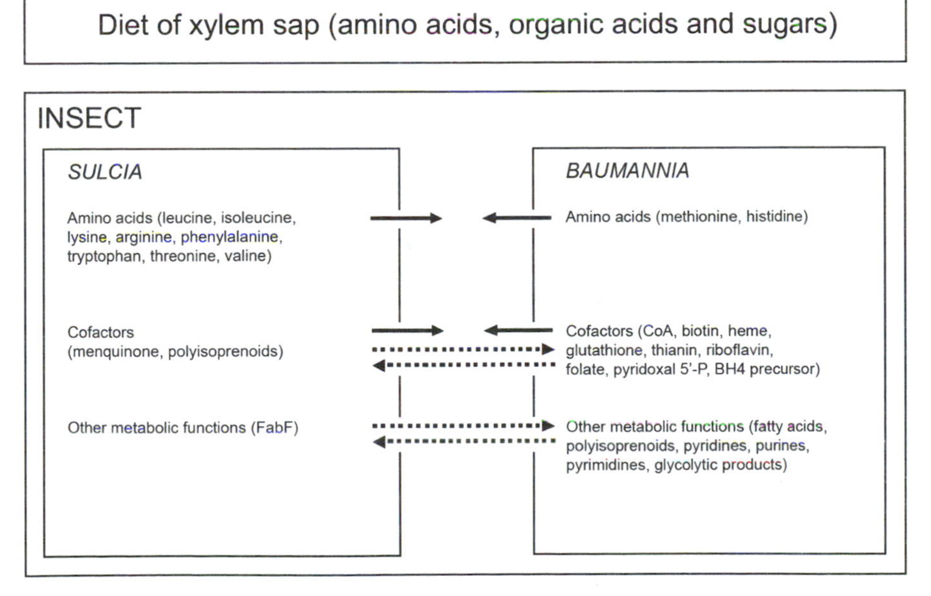

Figure 4-7 The chief metabolic capabilities of the symbiotic bacteria *Sulcia* and *Baumannia* in the insect host *Homalodisca coagulata*, as predicted from genomic sequencing. Solid arrows indicate nutrients transferred to insect, and dashed arrows indicate metabolites shared between the symbiotic bacteria. [Redrawn from figure 3 of McCutcheon and Moran (2007). Copyright (2007). National Academy of Sciences, U.S.A.]

examples of consortia.) A further animal apparently dependent on a microbial consortium is the oligochaete annelid *Olavius algarvaensis*. In an elegant genomic analysis, Woyke et al. (2006) identified two taxa of both sulfide-oxidizing and sulfate-reducing bacteria that probably form a sulfur-cycling consortium (see section 3.2.3) within their common host. Microbial consortia are also important among the gut microbiota of mammals. *Bacteroides* species in the human colon degrade complex polysaccharides to sugars (Xu et al. 2007), which are respired by *Bifidobacterium* and other anaerobic bacteria to lactate. The lactate is then fermented by bacteria such as *Eubacterium hallii*, producing butyrate, which is the principal respiratory substrate used by the gut epithelial cells (Flint et al. 2007).

In these various examples, the benefits derived from the different partners are either different (e.g., protective functions of the two snail species associated with *Chondrus crispus*) or a combined function (e.g., methionine derived by sharpshooter insects from their bacterial consortium). In other associations, the multiple partners may be closely related and their traits only subtly different. It can be argued that associating with multiple partners with similar traits can be advantageous because a mixed portfolio of partners would ensure advantage under a range of circumstances. As one putative example, various ascidians bear multiple genotypes of the cyanobacterium *Prochloron* that produce cyclic peptides called patellamides coded by the gene *patE* (Donia et al. 2006). The patellamides reach high concentrations in the symbiosis, often exceeding 0.1% of the total dry weight, and they are believed to protect the ascidian from predators. The *patE* sequences vary among the *Prochloron* genotypes, and therefore the total patellamide profile produced in a host is large and complex. The chemical complexity of the patellamides is believed to enhance the protective value of the defense. Intriguingly, the Tragedy of the Commons and Common Action models outlined earlier (section 4.4.1) would suggest that this symbiosis is susceptible to declining host benefit caused either by intersymbiont competition over host resources or by relaxed selection on patellamide production by each symbiont genotype. As this example illustrates, there is a need for definitive, empirical tests of the predicted consequences of mixed infections.

4.5 Living with the Chosen

Symbiosis researchers, like romantic novelists, tend to focus on the routes by which the partners come together, and to neglect the processes by which partners coexist in the long term. This is despite the fact that persistence is a defining characteristic of symbiosis.

This comment notwithstanding, I anticipate a renaissance of interest in "living with the chosen" in the coming years, following the recent upsurge of research on the defense systems of animals and plants, especially the innate immune system of animals. As a result, I am writing this chapter at a time when our understanding of how symbioses persist is founded on a mix of a small number of elegant classical studies and a few recent analyses using advanced molecular and cell biological techniques. Most of the classical studies have focused on describing patterns in the abundance and distribution of symbionts in hosts, and how these patterns are integrated into the developmental biology of their hosts; and I consider these in sections 4.5.1 and 4.5.2. I then address the interface between symbiosis and host defenses, especially in relation to the immune system (section 4.5.3).

4.5.1 Controls over Abundance and Distribution

In host-symbiont associations, the host, being larger, generally has a lower intrinsic growth rate than its symbionts. The persistence of the symbiosis depends on the host capacity to control the abundance and distribution of the chosen partners. The usual modes of control are restrictions on the growth and proliferation of the partners and on the amount of habitat space available to the partners.

Host controls over the location of its partners are the norm. Hollow thorns are the sole sites on South American *Acacia* trees suitable as nest sites for their ant partners (figure 1-2), and consequently the number of resident ants patroling the plant is limited by the number and size of the thorns. Among symbioses involving microorganisms, the rhizobia bacteria are restricted to root nodules of leguminous plants, *Chlorella* algae occur only in the digestive cells of hydra, *Buchnera* bacteria are located only in mycetocytes of aphids, the algae in many lichens occur as a defined layer between the upper cortex and lower medulla of the fungal thallus, and so on. Even the bacterium reported to be present in every cell of certain leaf-hopper insects, including the ommatidial cells of the eyes, is restricted to the cytoplasmic compartment of the insect cells (Nault and Rodriguez 1985).

The proximate determinants of localization are known for some symbioses. The *Chlorella* symbionts of hydra are phagocytosed at the apical surface of the host digestive cells and then transported on microtubules to their final location at the base of the digestive cell (Cooper and Margulis 1977). The underlying mechanism has not been established, but it might parallel the enhanced function of the microtubular motor, dynein, that mediates the intracellular movement of the pathogen *Campylobacter jejuni* (Hu and Kopecko 1999). The ordered spatial distribution of the lichen alga *Trebouxia* in foliose lichens such as *Parmelia* species is mediated by the growth and branching pattern of the fungal hyphae, each of which bears a single algal cell at its growing tip (Honegger 1984). The marine nematode worm *Laxus oneistus* bears ectosymbiotic sulfur-oxidizing bacteria on the posterior body wall, and this is dictated by the distribution of a host lectin which binds the bacteria (Bulgheresi et al. 2006).

The growth and division rates of the symbionts are strictly controlled to match that of the host in many symbioses, including the *Chlorella* in hydra (McAuley 1986), the bacterium *Buchnera* in aphids (Whitehead and Douglas 1993), and *Blattabacterium* in cockroaches (Brooks and Richards 1955). Symbiont abundance is regulated according to host demand for symbiotic services in plant root–microbial associations, with the level of mycorrhizal fungal infections depressed in high-phosphorus soils and the number of nodules on legume roots depressed in high-nitrogen soils (Smith and Read 2007; Kiers and van der Heijden 2006). The level of jasmonic acid in the plant has been implicated in the regulation of the mycorrhizal symbiosis (Hause et al. 2007). For example, infection of *Medicago truncatula* by AM fungi results in a two- to threefold increase in jasmonic acid titers, and plants engineered to have either reduced or increased capacity to synthesize jasmonic acid display delayed fungal colonization and reduced fungal abundance.

In comparison to the wealth of information on the patterns of symbiont abundance and distribution, very little is known about the mechanisms by which a host regulates symbiont growth and proliferation. The patterns of regulation have frequently been explained unsatisfactorily in terms of undefined growth factors and division factors. Genetic approaches are valuable to address this issue, as is illustrated by the analysis of Tanaka et al. (2006a) on the endophytic fungus in perennial ryegrass *Lolium perenne*. I have already introduced this study in chapter 2 (section 2.2.1) in the context of the evolutionary relationship between pathogens and mutualists. To recap, when a null mutation was introduced into the gene *noxA*, the fungal biomass increased dramatically, and the grass host grew slowly and senesced prematurely relative to the wild type. *NoxA* codes for NADPH oxidase, which generates hydrogen peroxide (H_2O_2) and other reactive oxygen species (ROS) by the reaction

$$\text{NADPH} + O_2 \rightarrow \text{NADP} + H_2O_2.$$

The ROS are hypothesized to activate signaling pathways in the fungus that regulate cell proliferation. Perhaps the host controls over fungal growth are mediated via this putative ROS-responsive pathway in the fungus.

One particularly intriguing instance of host control relates to the lifespan of arbuscules, the dichotomously branching haustoria of arbuscular mycorrhizal fungi that are the site of inorganic phosphate transfer from the fungus to plant. The arbuscule is a short-lived structure, usually persisting for 7–10 days; and the plant cell accommodating an arbuscule survives the death of the arbuscule. As a result, each root bears multiple arbuscules at various stages from initiation to functional maturity to degeneration and death. The plant plays an important role in determining the lifespan of each arbuscule, linked to its nutritional function. This is indicated by experiments of Javot et al. (2007) on a mutant of the plant *Medicago truncatula* that lacked the phosphate transporter expressed only at the plant-arbuscule interface. The arbuscules in these mutant plants developed but then died prematurely, their lifespan reduced from an average of 8.5 days for the wild type to 2–3 days. The fungus failed to proliferate in the root and the symbiosis was terminated.

In some symbioses, the host does not appear to maintain precise control over symbiont proliferation and the excess symbionts are regularly shed from these symbioses. A spectacular example of regulation by elimination is provided by the relationship between the squid *Euprymna scolopes* and its complement of luminescent *Vibrio fischeri*. At night, the bacteria are densely packed in tubules of the squid light organ at concentrations up to 10^{11} cells ml^{-1}, where they luminesce. At dawn, ~90% of the cells are shed from the light organ into the water column, and the remainder proliferate through the day (Ruby and Asato 1993). For the host, the discarded microorganisms are wasted resources because they contain host-derived nutrients. In the case of the *E. scolopes* symbiosis, the shed symbionts saturate the immediate environment (Lee and Ruby 1994), and this would promote access of host offspring to a source of compatible symbionts, but no quantitative analysis of the nutritional/energetic cost and reproductive benefit arising from symbiont expulsion has been conducted.

For a minority of symbioses, there is no opportunity for coordination of the growth of the partners. The partners in most associations between hermit crabs and sea anemones are linked by the empty shell of a gastropod mollusk; the *Pagarus* crabs colonize the interior of empty gastropod shells, while the sea anemones use the shell surface as a substratum. Until the crab grows too large to fit into the shell, it benefits from the extra protection from predators afforded by the sea anemone, while the sea anemone gains discarded food items from the crab. When the hermit crab has outgrown the shell, it abandons the sea anemone partner and is very vulnerable to predators while finding a larger shell (with the additional potential cost that the new shell may lack a

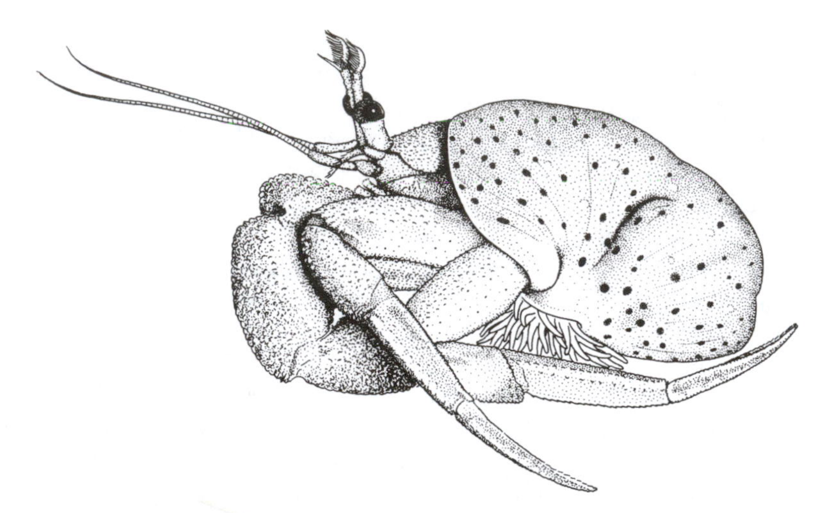

Figure 4-8 The symbiosis between *Pagarus prideauxi* and *Adamsia carciniopados.* [Reproduced from figure 62 of Manuel (1988)]

sea anemone). One species, *Pagarus prideauxi*, has overcome this limitation through its specific association with the sea anemone *Adamsia carciniopados* (figure 4-8). Unlike most *Pagarus*–sea anemone symbioses in which the sea anemone mounts the shell, *P. prideauxi* takes the initiative by grasping an individual *Adamsia* with its claws and inserting the anemone on its ventral surface. In this location, the anemone is unable to provide much protection against predators in the short term but it grows rapidly, being well positioned to access the crab's food. Its growth pattern is most unusual, involving the rapid expansion of the pedal disk, which grows laterally around the body of the crab as a horny sheet, ultimately to form a continuous layer over the dorsal surface of the crab. The *Adamsia* pedal disk is a living and growing extension of the gastropod shell. As a result, the crab does not outgrow its "symbiotic shell" and the symbiosis can persist for the full lifespan of the hermit crab.

In complex hosts, living with the chosen involves not only the balanced growth rates of the partners but also integration of symbionts into the developmental program of the host. This issue is addressed next.

4.5.2 *Integration into the Developmental Program of the Host*

An elegant demonstration of symbiont integration into host development is provided by the classic research of Cleveland and colleagues (Cleveland et al. 1960) on a group of cockroaches, the woodroaches of the family Cryptocercidae. Anaerobic cellulose-degrading protists in the anoxic hindguts of *Cryptocercus* species enable these insects to subsist on their diet of wood. The metabolic end-products of the microbial cellulose digestion and subsequent anaerobic respiration are short-chain fatty acids, which are assimilated across the hindgut wall and utilized as carbon substrates for aerobic respiration of the insect. This relationship involves asexually reproducing protists, which proliferate by binary fission and are extremely sensitive to oxygen. When the insect host molts, the cuticular lining of the hindgut is lost and the hindgut contents become oxygenated. The protist symbionts would be killed if they did not undergo a developmental change from the asexual to the sexual form, a process that involves the formation of oxygen-resistant cysts. The encysted protists are expelled from the insect in fecal pellets and, after the insects have molted, they feed on these pellets and reestablish the symbiosis.

The coordination of protist and insect development, such that the protists encyst prior to the host molt, is controlled by the insect ecdysteroid molting hormone, with both cyst formation and molting orchestrated by elevated hormone titers. Sexual cysts are induced experimentally by injecting synthetic ecdysteroid into intermolt roaches and can be prevented by extirpation of the insect neurosecretory cells which produce the ecdysteroid hormone (Cleveland et al. 1960).

The developmental program of shedding and resynthesis of the gut symbioses in woodroaches is unusual. Many symbioses are retained through developmental changes of the host, and host-symbiont integration is particularly important for vertically transmitted symbioses which, by definition, persist through the lifecycle of the host. Diverse routes of integration have evolved among the vertically transmitted mycetocyte symbioses of insects, often involving complex patterns of changes in localization and abundance of symbionts linked to host developmental age (Buchner 1965). A key challenge is to understand the molecular basis of these patterns. Of particular interest, the mycetocytes of aphids express homologs of key developmental genes, including *distalless*, otherwise implicated in limb development, and *engrailed*, generally required for formation of posterior compartments (Braendle et al. 2003). These insect genes have presumably been recruited to a novel role in the symbiosis. The implication is that signaling networks controlling the integration of the symbiosis into the wider developmental program of the host may have involved the remodeling of preexisting networks and gene function, rather than the evolution of novel systems and genes unique to the symbiosis.

Consistent with this conclusion, global analyses of gene expression in various symbioses have generally demonstrated that symbiosis affects the expression of genes with homologs of known function found in nonsymbiotic organisms and have not implicated substantial numbers of novel genes in symbiosis function. Microarray analyses of plants infected with ectomycorrhizal fungi have revealed the upregulation of genes involved in a variety of cellular functions, including fungal cell wall synthesis, cell division and proliferation, and plant metabolism (especially glycolysis and amino acid synthesis) and response to stress (Martin et al. 2007). A comparison of the transcript profiles of the sea anemone *Anthopleura elegantissima* naturally bearing and lacking the symbiotic alga *Symbiodinium* identified differences in the expression of genes with roles in lipid metabolism, cell proliferation and adhesion, and responses to stress (Rodriguez-Lanetty et al. 2006). Similarly, many of the genes of leguminous plants expressed exclusively in root nodules code for previously described functions, including channel and transporter proteins and metabolic enzymes (Jones et al. 2007).

4.5.3 Symbiosis and Host Defenses

The interaction between symbioses and host defenses is crucial to our understanding of how the chosen are controlled. Defense systems are essential because every organism is a concentrated source of nutrients that, unless protected, is swiftly invaded and consumed by other organisms. The efficacy of the defense systems is demonstrated by the far greater diversity of saprotrophs consuming a cadaver than parasites exploiting an equivalent living organism. The defense systems are complex and varied principally to combat the diverse and ever-changing challenges posed by parasites and pathogens. Nevertheless, effective defenses are a balancing act because they must both be compatible with the persistence of beneficial partners and avoid self-inflicted damage to the host. Indeed, various symbionts protect their habitat (i.e., the host) against invasion by competitors, including parasites, and, thereby, can make a valuable contribution to host defenses against parasites.

A key feature of host defenses is that they can be involved in host interactions with both symbionts and pathogens. This has been revealed by a study of the bacterium *Aeromonas veronii*, which is a gut symbiont required by the medicinal leech *Hirudo verbena* (=*H. medicinalis*), probably because it provides vitamins that supplement the host diet of blood; and it is also an opportunistic pathogen of mammals, including immunocompromised people. *A. veronii* has a type III secretion system (T3SS), functionally equivalent to that in rhizobia, discussed earlier in this chapter (see section 4.2.3). When the T3SS of *A. veronii* is compromised by mutation of an essential component, the bacterium displayed both reduced capacity to colonize the leech gut and reduced virulence in mice (Silver et al. 2007). In other words, the T3SS is required to maintain both symbiotic and antagonistic infections. Furthermore, the two outcomes have a common basis: the T3SS is involved in manipulating the host innate immune system, particularly preventing the phagocytosis and clearance of the bacterial cells by macrophages. The components of T3SSs and their effector proteins are often referred to as virulence factors because they mediate interactions between various bacterial pathogens and eukaryotic cells. As the function of T3SSs in *A. veronii* and rhizobia illustrates, this term can be misleading; the T3SS and its effector proteins promote bacterial colonization, which may—and may not—be deleterious to the host.

The traditional explanation for the persistence of symbioses is that an acceptable symbiont is one that can circumvent host defenses, either by living in an immunologically privileged site or by specific adaptations to evade or suppress recognition by the immune system. This type of explanation is simplistic. Recent developments in the fields of immunology and symbiosis research have revealed that the interactions between symbionts and the defense systems of their hosts can be much more complex than simple evasion.

Let us, first, consider the significance for symbioses of immunologically privileged sites, i.e., locations in the body that effectors of the immune system either cannot access or where they are suppressed. The principal sites in vertebrates are quoted as the brain, anterior chamber of the eye, testes, and cell contents (Henderson et al. 1999); and none are exploited by symbionts, contrary to the expectation that they would be colonized preferentially. In fact, colonization by symbiotic microorganisms is particularly prevalent in one of the immunologically most active sites, the gastrointestinal tract.

The intracellular environment deserves further consideration. It is immunologically privileged only in the narrow sense of being protected from the cellular and humoral defenses of the extracellular compartment (e.g., animal blood, plant apoplast). Cells are well defended against colonization. As discussed in chapter 2 (section 2.5), the adaptive immune system of vertebrates poses barriers to the colonization of cells that have been overcome only by a few microbial pathogens and apparently no symbionts. The innate immune system also contributes to the within-cell defense, mediated by intracellular receptors, including the NOD proteins in mammalian cells and PGRP-LC (peptidoglycan recognition protein-LC) in insects. For example, when PGRP-LC of *Drosophila* binds to bacterial peptidoglycan, the Imd signaling cascade is triggered, leading to the activation of the NF-κB transcription factor Relish, which mediates the upregulation of genes coding antimicrobial peptides and other defensive responses.

Consistent with these indications that intracellular symbionts are not sequestered away from the host immune system, there is evidence that the persistence of these symbionts depends on a dynamic interaction with subcellular defenses of the host. Intracellular symbionts generally gain access to host cells by phagocytosis. Phagosomes are, by default, trafficked to the lysosome, where the phagosomal cargo is degraded by a suite of acid hydrolases, including proteases, glycosidases, and lipases. For a symbiont to evade this fate, the standard processing of phagosomes must be modified. Ultrastructural studies suggest that a minority of symbionts escape from the host membrane and lie free in the cytoplasm; these include *Wigglesworthia* in tsetse fly mycetocytes. Nothing is known about the underlying mechanisms, beyond the absence of evidence for any genes in the *Wigglesworthia* genome that might function as lysins (proteins that mediate the release of some pathogens, such as *Listeria*, into the cytoplasm). Most symbionts remain within the host membrane. In the few systems examined, the lipid and protein content of this host membrane is very different from the cell membrane and, in recognition of these changes, the membrane is known as the symbiosomal membrane and the contents as the symbiosome (Roth et al. 1988). Components on the symbiosomal membrane are predicted to prevent the trafficking of the symbiosome to the lysosome. Of particular interest are the Rab proteins, a family of small proteins with GTPase activity that play a key role in the processing of vesicles. In particular, Rab5 is a signature protein of early endocytic vesicles and Rab7, which becomes associated with the vesicle membrane at a later stage in vesicle maturation, promotes lysosomal fusion. An analysis of the distribution of these proteins in the sea anemone *Aiptasia pulchella* revealed that Rab5, and not Rab7, was localized to most symbiosomal membranes bounding the algal symbiont *Symbiodinium* (Chen et al. 2004). This suggests that phagosomal maturation is arrested by *Symbiodinium*. The underlying mechanisms are not known, but they may parallel a similar capability displayed by some facultative intracellular parasites, e.g., *Mycobacterium*, *Leishmania* (Knodler et al. 2001). For the *Symbiodinium* system, most of the symbiosomes lose Rab5 and become Rab7-positive, presumably destined for lysosomal fusion when animals are incubated with the photosynthetic inhibitor DCMU (Chen et al. 2004). These data suggest that the identity of Rab proteins delivered to the symbiosomal membrane is determined by the photosynthetic competence of the algal symbionts.

Evasion of the host defenses is also important for persistence of microorganisms in the extracellular environment. Antigenic variation is an important route by which a number of parasites circumvent the adaptive immune system of vertebrates, but this trait is not known among symbionts. Some pathogens display molecular mimicry of their host. For example the dominant carbohydrate residue on surface glycoproteins and glycolipids of the pathogen *Haemophilus influenzae* is sialic acid, as also used by mammalian cells, and not the mannose residues found on most bacteria. The adoption of fucose as a surface sugar moiety on the gut symbiont *Bacteroides fragilis*, as discussed earlier in this chapter (section 4.4.1), may play a similar role.

Despite these specific examples, it is improbable that symbiont persistence depends exclusively on symbiont adaptations to evade host defenses. It is in the interests of the host to accommodate symbionts and some selectivity in the defense system is expected, to discriminate between symbionts and potentially deleterious colonists. This expectation is not readily compatible with traditional self-nonself models of immune function. By these models, the adaptive immune system of vertebrates is considered to be an exquisitely specific system for detecting and eliminating any entity with molecular patterns different from those that the immune system has been educated to identify as self; while the innate immune system in all animals includes various pattern recognition receptors (PRRs) which initiate defense responses when they bind to common molecular features of microorganisms termed pathogen-associated molecular patterns (PAMPs) (Medzhitov and Janeway 2002). These models require complex explanations to account for the apparent ineffectiveness of the immune system in eliminating the resident microbiota of animals, including the estimated 90% of cells in the human body that are nonself.

The expectation that the immune system can discriminate between deleterious and harmless or beneficial partners is addressed by an alternative model of immune function, known as the danger model (Matzinger 2002). The danger model is founded on the notion that the immune system is activated by danger/alarm signals produced by injured cells. Seong and Matzinger (2004) note that hydrophobic portions of molecules are generally buried within macromolecules or membranes but become exposed on damage, for example, through protein misfolding or disruption to the membrane or extracellular matrix of cells. They call these hydrophobic motifs DAMPs (danger-associated molecular patterns), and argue that the immune system is built on an ancient, possibly universal, damage-protection system. The danger model predicts that symbioses persist when they are insufficiently dangerous to trigger the immune system.

Some data suggest that the immune system is more than a guardian, and that it also contributes to the management of microorganisms. To illustrate, let us consider a specific molecule in the innate immune system of *Drosophila*. PGRP-LB is an effector of the Imd pathway of insects,

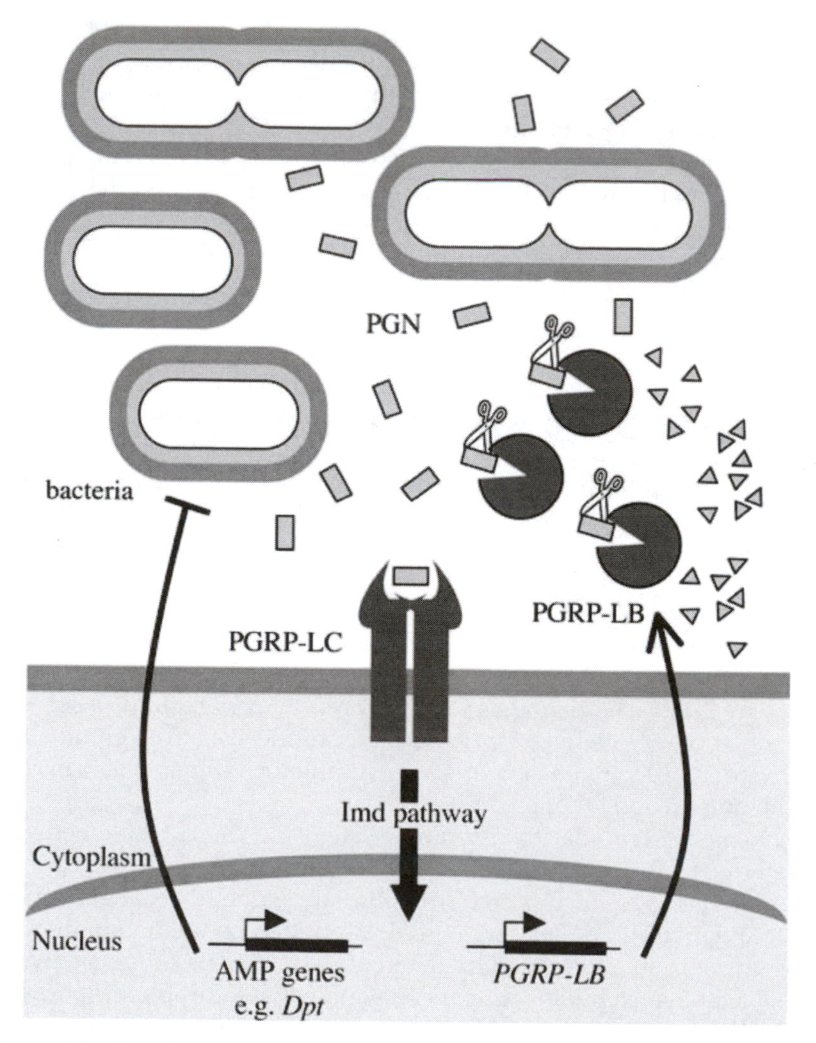

Figure 4-9 Regulation of the insect Imd (immune defence) pathway by PGRP-LB (peptidoglycan recognition protein-LB). Peptidoglycan (PGN) released from bacteria activates the Imd pathway by binding to the receptor PGRP-LC, resulting in the induction of genes for both antimicrobial peptides (AMP), such as diptericin (Dpt), and PGRP-LB, a PGN-amidase (illustrated as scissors transforming PGN ■ to inactive fragments ▲). By eliminating PGN, PGRP-LB downregulates the Imd pathway. In this way, the amount of PGRP-LB produced can regulate the immunoreactivity of the cell. [Reproduced from *Immunity*, figure 7 of Zaidman-Rémy et al. (2006) with permission from Elsevier]

which is activated by the binding of bacterial peptidoglycan to a cell membrane receptor, PGRP-LC (figure 4-9). Unlike PGRP-LC and most other PGRPs, PGRP-LB is enzymatically active, an amidase which degrades the peptidoglycan that it binds (Zaidman-Rémy et al. 2006). By eliminating peptidoglycan, PGRP-LB downregulates the Imd pathway. The other genes activated by the Imd pathway are antimicrobial peptides. The amount of PGRP-LB produced per peptidoglycan molecule binding to PGRP-LC determines the antimicrobial response of the cell to the peptidoglycan inducer and, ultimately, the immunoreactivity of the host cell or organ. PGRP-LB production is predicted to be higher in cells tolerant of microbial interactions than in intolerant cells. Consistent with this expectation, PGRP-LB transcripts are enriched in mycetocytes relative to other organs of the weevil *Sitophilus zeamais* (Anselme et al. 2006).

The mode of action of PGRP-LB suggests that the persistence of symbionts depends on their engagement with and management by the host immune system. If the immune system is a manager of interactions, then the defensive response against overt pathogens is just one aspect of immune system function, and both the self-nonself and danger models define immune system function too narrowly. Relevant to this interpretation, the selection pressure to manage a complex microbiota has been invoked as a contributory factor for the evolution of the vertebrate adaptive immune system (McFall-Ngai 2007).

4.6 Résumé

Symbiotic organisms are selective in their choice of partners. In some systems, this selectivity can be explained in terms of single, defining traits, such as the dimensions of the entrance hole to domatia in plant hosts of protective ant colonies. For other symbioses, partner choice is mediated by multiple, complex signaling interactions (e.g., the synthesis of the legume-rhizobium symbiosis involves the flavonoid-Nod–factor-kinase exchange, lectin-carbohydrate interactions, plant cytokinin pathway, and various T3SS effector proteins of the rhizobia). A plausible interpretation of the complex interactions is that these symbiotic organisms use multiple different criteria to assess the suitability of potential partners. The relative importance of the different criteria may depend on the range of partners available, environmental circumstance, and the incidence of cheaters and other deleterious organisms.

Symbiotic organisms vary widely in their specificity, i.e., the range of partners with which they associate. The availability of suitable partners is recognized as an important determinant of specificity: organisms with readily available potential partners can afford to be more choosy than those that rarely encounter compatible partners; while the combination of low partner availability and high incidence of cheaters can select for targeted transmission mechanisms that provide for both high specificity and assured access to partners. Low specificity can lead to mixed infections, involving multiple genotypes with similar traits. This condition is predicted to be deleterious for the host, because the taxa either compete over host resources or become progressively less beneficial by failing to contribute their "fair share" of services to the host. These predictions do not apply to the various associations where the multiple partners either confer distinct services, such as protection against different natural enemies, or are mutually dependent, as in certain animal symbioses with microbial consortia. Further research on the incidence and consequences of multiple closely related symbiont taxa in individual hosts is needed. This will be facilitated by recent advances in sequencing methods, enabling the genetic diversity of organisms in symbioses to be assessed comprehensively.

Our increasing understanding of the processes involved in the formation and persistence of symbioses can shed light on the evolution of symbioses. One striking theme emerging from studies on a variety of associations is that the mechanisms underlying symbioses commonly involve preexisting capabilities, including the recruitment of gene products with nonsymbiotic functions to novel symbiosis-specific functions (section 4.5.2). These results raise the question of how the expression of these many and diverse genes is regulated in the context of the symbiosis. Should we be looking to a single master gene that coordinates the network of relevant genes in the symbiotic context, or is control distributed across many preexisting regulatory circuits? There is also evidence for similarities in the mechanisms of symbiosis formation among different associations. In particular, a common signaling pathway, the sym pathway, is harnessed by various mutualists and parasites to gain entry to plant roots (section 4.2.4), and certain beneficial and pathogenic microorganisms manipulate the endosomal system of host cells in the same way (section 4.5.3). Single processes, once evolved, may be used, reused, and reworked in both antagonistic and mutually beneficial interactions.

CHAPTER 5

The Success of Symbiosis

IN THE OPENING CHAPTER of the book, I described the evidence for the biological significance of symbiosis. In particular, many symbioses are biologically successful, in that they are evolutionarily persistent, widespread in the environment, and ecologically important (see especially chapter 1, section 1.4). Here, I revisit this general theme with the purpose of using concepts and conclusions obtained in chapters 2–4 on the formation and persistence of the symbiotic habit to address the processes underlying the success of symbioses.

The first sections of this chapter concern two traditional indices of biological success: the impact of symbiosis on the evolutionary diversification of organisms (section 5.2) and on the structure of ecological communities (section 5.3). I then turn to address the success of symbioses in the context of human activities. Because of their traits, symbiotic organisms have been argued to be particularly vulnerable to environmental anthropogenic effects and, contrarily, of great potential value for exploitation to human benefit. For example, symbioses, including ecologically important associations, may be predisposed to extinction because all members of an association are threatened by the loss of one partner; and many symbiotic microorganisms have the life-history traits of slow growth rates in highly structured environments, making them likely sources of bioactive compounds with uses as novel therapeutic or pest control agents. I consider the success of symbioses in the context of extinction risk, climate change, and biotic homogenization in section 5.4; and the opportunities to use our understanding of the formation and persistence of symbiosis to promote human health and food production in section 5.5.

5.2 SYMBIOSIS AND THE EVOLUTIONARY DIVERSIFICATION OF ORGANISMS

5.2.1 The Diversification of Symbiotic Organisms: Ants and Corals

As stated in chapter 1 (section 1.4.2), various organisms have gained access to novel traits by forming symbioses. In particular, they have acquired metabolic capabilities, chemical or physical protection from natural enemies, and mobility (table 1-1). Symbiosis is, in this way, an evolutionary innovation that can make new niches available by allowing organisms to exploit resources underutilized by other taxa, escape from predators, and so on. This ecological opportunity can trigger the diversification of the symbiotic organisms.

There is now persuasive evidence that the diversification of several major groups has been founded on symbioses. The acquisition of bacteria that gave rise to mitochondria allowed eukaryotes to escape from anoxic environments, and the resultant diversification comprises many groups of exclusively or predominantly unicellular protists and three major multicellular radiations (the animals, fungi, and plants) (Baldauf 2003). The mycorrhizal symbiosis was the key innovation permitting plant acquisition of nutrients from the substratum and leading to the transition to land and subsequent diversification of terrestrial plants (Pirozynski and Malloch 1975; Simon et al. 1993). Among terrestrial vertebrates, herbivory has evolved at least 50 times over the last 300 million years, invariably from carnivorous or insectivorous ancestors and in most instances underpinned by symbiosis with cellulolose-degrading microorganisms (Vermeij 2004). Mutualisms have also promoted the diversification of taxa at the phylogenetic levels of order to family. Research on relationships involving two groups, the ants and scleractinian corals, has provided valuable insight into the underlying processes.

The diversification of ants has been addressed by Wilson and Hölldobler (2005). Ants probably evolved in the early Cretaceous, ~115–35 million years ago (Brady et al. 2006), with a likely ancestral lifestyle as a generalist predator on the forest floor. About 40–50 million years ago, some ants, especially members of the Formicinae and Dolichoderinae, adopted an arboreal habit, where they diversified. This change in lifestyle was underpinned by a shift in the ants' dietary habits to dependence on the extrafloral nectar of plants and the honeydew of hemipterans. [Honeydew is essentially the phloem sap of plants, albeit somewhat modified by passage through the hemipteran gut (Douglas 2006).] The dietary change arises from the adoption of mutually beneficial relationships based on the reciprocal exchange of food for the ants and protection for the hemipterans or plants. Essentially, mutualisms have transformed ants into herbivores that consume plant fluids. This is revealed by stable isotope analysis. The value of $\delta^{15}N$ (^{15}N:^{14}N) increases through the trophic levels from plant through herbivore to carnivore because the ^{14}N isotope is preferentially lost in catabolic reactions (and ^{15}N retained). Davidson et al. (2003) found that, for arboreal ants in tropical forests of both Peru and Brunei, the $\delta^{15}N$ of ants tending hemipterans was similar to that of insect herbivores, and lower than that in predatory ants (figure 5-1).

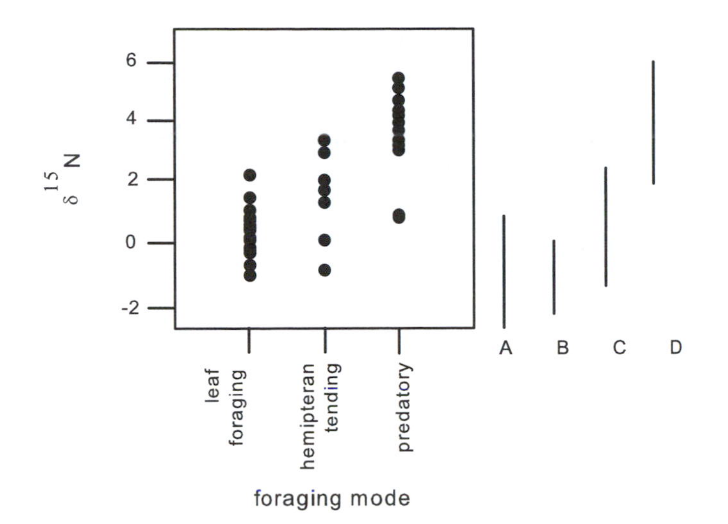

Figure 5-1 Mean $\delta^{15}N$ values (‰) of ant species in a Borneo rainforest, classified according to predominant feeding mode and calibrated against the range of $\delta^{15}N$ of plants (A), hemipterans (B), chewing herbivorous insects (C), and arthropod predators (D). [Reproduced from *Science*, figure 2 of Davidson et al. (2003), omitting species with mixed or uncertain foraging strategies. Reprinted with permission from AAAS]

The evolutionary importance of mutualisms for ants and their partners extends beyond the food-for-protection relationships described in the preceding paragraph. In some ant-hemipteran associations, the ants confer both protection and mobility to their insect partners (see table 1-1). The ants transport individual hemipterans to plants, on which the hemipterans feed, producing the honeydew that the ants consume. This is advantageous to the small hemipteran insects, for which locating a host plant is otherwise a risky activity, and it also promotes honeydew supply for the ants. Of particular note, the mealybug *Malaicoccus formicarii* in Malaysia is transported daily in the mandibles of its associated ants *Hypolcinea cuspidus* from the nest to plants in the morning and back to the nest in the evening. This relationship is most remarkable in that the nest is composed entirely of living bodies of the ants (Maschwitz and Hanel 1985). The ants cling together between two leaves, forming a living chamber within which the mealybugs are retained. The location of this living nest is changed every few days to weeks, according to the local availability of suitable plants for the mealybugs. It appears that the ants live exclusively on the honeydew of the mealybugs. This relationship is reminiscent of human pastoralist nomads who shift around, following the availability of suitable pasture for their domestic animals.

Ant agriculture is also displayed in the cultivation of fungi. This symbiosis has evolved in a single New World tribe of myrmecine ants, the Attini, including the leaf-cutter ants, e.g., *Atta* and *Acromyrmex*. These insects harvest plant material, as if they were herbivores, but do not eat it. Instead, they chew and bite the food and then transfer it to their fungal symbiont maintained in the nest. The plant material is decomposed by the fungus, and the ants feed on the fungus. This symbiosis is immensely successful, being the single most important herbivores in various savannah and rainforest habitats, although their diversity, comprising some 200 species, is modest by the standards of other herbivorous insect groups (Hölldobler and Wilson 1990).

The diversification of ground-dwelling ants has also been promoted by a mutualism, in this case the ant-mediated dispersal of plant seeds, especially those bearing elaiosomes (Wilson and Hölldobler 2005). Elaiosomes are lipid-rich food-bodies attached to the seeds of certain plants. Some ant species harvest the seeds and consume the elaiosome, usually leaving the dispersed seed intact. The inclusion of elaiosomes in the diet has increased the ecological range of ants from the forest floor to grassland and other dry habitats, especially in Australia, South Africa, and Mediterranean countries; and the ants have diversified in these different types of vegetation. These associations are generally not symbioses (the duration of contact is too brief) but the relationships of some arboreal ant species of the genus *Azteca* and plants in Central America are more persistent. The *Azteca* ants harvest seeds, particularly of epiphytic plants, and are reported to sow them specifically in their nests. The resultant roots and stems of the epiphytes provide a structural support for the ant nests, and the entire structures are known as ant gardens (Kaufman and Maschwitz 2006). In this unusual symbiosis, ant-conferred mobility is exchanged for a stable nest site.

There are also instances of enhanced diversification of some mutualistic partners of ants, particularly among the relationships with plants that are based not on ant-mediated protection, but on their provision of nutrients. These associations are restricted to epiphytes (i.e., plants that live on other plants from which they do not derive food), and have evolved in at least seven plant families. The key plant trait is an expanded, hollow chamber in the stem or leaves, within which the ants nest. Nutrients from material ants bring to their nest are absorbed across the chamber wall into the plant tissues, complementing the nutrients the epiphytes derive from rainwater. In this way, these ant-associated plants are released from the ecological requirement for high rainfall, enabling them to diversify from rainforests into grassland and other habitats (Huxley 1980).

In addition to the five types of mutualism described above, ants associate with lycaenid butterflies as I have discussed in chapter 2 (section 2.2.5). Why are ants so predisposed to mutualisms, including symbioses, with such dramatic consequences for their diversification? The ancestral ants were mobile, generalist scavengers foraging on the ground and vegetation, with most individuals unable to fly. These insects, therefore, repeatedly came into contact with plants, hemipterans, and fungi while searching and sampling for food and nest sites. Because the plant, fungus, or insect is not interacting with a single ant but with a colony of many ants, the scale of mutual benefit can be substantial from the initial interactions. In this way, the sociality of ants has also contributed to the propensity of ants for mutualistic interactions.

In summary, the ancestral ant traits of sociality, ground-dwelling lifestyle, and generalist feeding habits underpin their propensity for mutually beneficial interactions, including symbioses; and these interactions have promoted the diversification of ants and, in some systems, their partners.

My second example of symbiosis-dependent diversification concerns the scleractinian corals, and I focus on the symbiosis with photosynthetic algae. Approximately half of the ~1300 known scleractinian species bear the dinoflagellate alga *Symbiodinium*. All scleractinian corals have a calcareous exoskeleton and, when the coral skeleton is produced more rapidly than it is eroded, reefs are generated. (Reefs are defined formally as wave-resistant biotic structures.) Today, the symbiotic scleractinian corals are the dominant reef-builders in shallow waters at low latitudes and the nonsymbiotic corals generate reefs at the lower reaches of the continental shelf. There is unambiguous fossil evidence for scleractinians in the mid-Triassic, about 240 million years ago. These forms were neither abundant nor reef-building, and they are believed not to have borne symbiotic algae. Scleractinians subsequently diversified, with particularly high rates of origination in the Norian-Rhaetian phases of the upper Triassic (figure 5-2). At this time, they were reef-building. Scleractinian diversity was dramatically curtailed by the mass extinction at the end of the Triassic but, after an interval of 5–8 million years of the early Jurassic, they recovered. The subsequent evolutionary history of the scleractinian corals through the Mesozoic and Cenozoic has been described by Stanley (2003) as "amazingly hearty. These corals survived devastating environmental catastrophes, including a number of mass extinctions."

The near-simultaneous diversification and onset of reef-building in the upper Triassic scleractinians have been attributed to the acquisition of symbiotic algae. Remarkably, there is chemical evidence that corals of the upper Triassic bore symbiotic algae. This comes from the isotopic

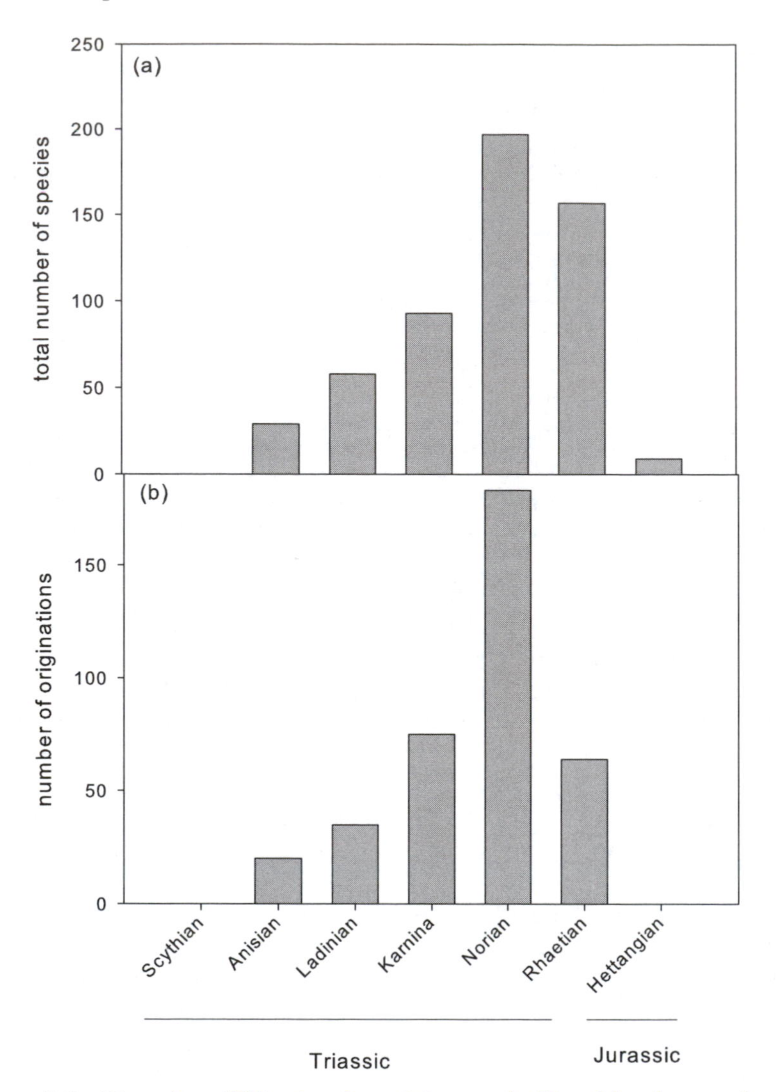

Figure 5-2 Diversity of Triassic scleractinian corals. The oldest known fossils are from the Anisian stage. The total diversity (a) and number of originations (b) are maximal in the Norian, when these corals are believed to have acquired the photosynthetic symbiosis [Redrawn from figure 2 of Stanley and Swart (1995)]

composition, specifically the $\delta^{13}C$ and $\delta^{18}O$, of the coral skeleton. The $\delta^{13}C$ and $\delta^{18}O$ values vary widely and are strongly positively correlated in the skeletons of modern nonsymbiotic corals, but are very uniform in symbiotic corals. Although the reasons for this difference between symbiotic and nonsymbiotic corals are not understood fully, their distinctive isotopic signatures can be used to infer the status of fossil skeletons. Unfortunately, fossil skeletal material is rarely pristine because the aragonite skeleton is not very stable and converts readily to calcite. Even so, Stanley and Swart (1995) were able to obtain data for some upper Triassic fossils, and these samples had an isotopic signature similar to modern symbiotic corals. In other words, there is evidence that the algal symbiosis evolved at the time when scleractinians diversified and initiated reef-building. It is inferred from the correlated changes that the acquisition of the symbiosis led to reef-building, allowing the corals to exploit resources in novel ways; and this wealth of ecological opportunities triggered an adaptive diversification of the corals.

A rational basis for the proposed causal relationship between the symbiosis and reef-building is provided by physiological data on modern coral symbioses. Specifically, corals display light-enhanced calcification, meaning that their deposition of calcareous skeleton is promoted by photosynthetic activity of the symbiotic algae (Goreau 1959). Although described some 50 years ago, the underlying mechanisms are still not understood well and are likely to be complex (Moya et al. 2006), since the site of calcification is physically remote from the algal cells. One idea is that photosynthesis promotes the supply of inorganic carbon from metabolic pools for calcification. Although this proposal is counterintuitive (photosynthesis consumes inorganic carbon), it is supported by experimental data. Specifically, light-enhanced calcification is depressed by inhibitors of both algal photosynthesis and carbonic anhydrase, the enzyme which accelerates the interconversion of CO_2 and HCO_3^-. Another potentially important effect is the net consumption of protons by photosynthesis. This could have a general effect on the acid/base balance of the organism, including alkalinization at the calcification site, which would drive the equation below toward carbonate production.

$$CO_2 + H_2O \leftrightarrow H_2CO_3 \leftrightarrow H^+ + HCO_3^- \leftrightarrow 2H^+ + CO_3^{2-}.$$

Photosynthesis could additionally influence the pattern of calcium ion transport, perhaps linked to its impact on carbon supply or acid/base balance.

Because the physiological basis of light-enhanced calcification in scleractinian corals is unresolved, we do not know if this effect is relevant to other animal calcification systems. This caveat is important because of current uncertainty about the role of photosynthetic symbioses in reef-building animals of the Paleozoic (before the evolution of scleractinians in the Mesozoic). Paleozoic fossils are generally not suitable for isotopic analysis and many studies have focused on skeletal characters. One of the more plausible Paleozoic photosymbioses is a group of Permian bivalves called the alatoconchids, which had huge dorsoventrally flattened extensions to their valves potentially suitable for light capture. The tabulate corals, which contributed to post-Cambrian Paleozoic reefs, have also been suggested to have borne photosymbionts (Coates and Jackson 1987), but this viewpoint is disputed (Wood 1997). Paleo-ecological studies of the environmental conditions in ancient reefs also suggest that most Paleozoic reef-builders were not photosynthetic (Wood 1998).

In summary, there is strong evidence that symbiosis has triggered the evolutionary diversification of ants and scleractinian corals by allowing these organisms to exploit their environment in different ways. These and other diversifications have, however, been underpinned by multiple traits of the organisms additional to symbiosis, for example the sociality of ants and coloniality of reef-building corals. Evolutionary diversification is not an inevitable consequence of symbiosis, and some symbiotic taxa are not speciose.

5.2.2 *Diversification through Coevolution and Cospeciation*

In the section above, I treated symbiosis simply as a route by which organisms gain access to novel capabilities or habitats. This is a simplification. The diversification of symbiotic organisms can also be promoted by evolutionary interactions with their partners. The key processes are coevolution, i.e., the reciprocal evolution of traits of both organisms in an association in response to traits of their partners, and cospeciation, the synchronized speciation of two organisms.

The classic example of rapid diversification of organisms in mutualisms through coevolution and cospeciation concerns plants and pollinating insects. This process is widely accepted to have triggered the adaptive radiation of angiosperms and insects in the Cretaceous through to the Tertiary. Angiosperm species pollinated by animals are more diverse than sister taxa pollinated by abiotic agents (Dodd et al. 1999). Considering one particular group of plants, the classic research of Verne and Karen Grant (1965) has revealed the evolution of the floral morphology of Polemoniaceae from an ancestral form pollinated by bees to a wide array of different forms, under selection from different pollinators, including butterflies, moths, beetles, hummingbirds, and bats (figure 5-3).

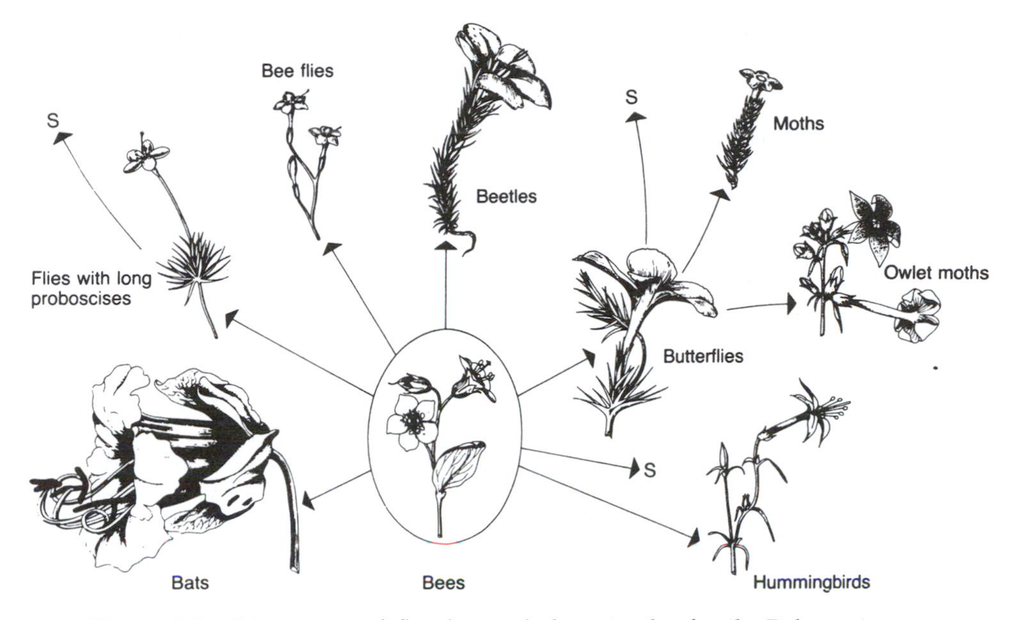

Figure 5-3 Divergence of floral morphology in the family Polemoniaceae under selection from different pollinators. Self-pollinating species (S) have evolved in various lineages. [Reproduced from figure 8-5 of Howe and Westley (1988)]

Let us consider cospeciation further. In principle, the partners in a symbiosis may cospeciate because of high specificity imposed by one (or both) of the partners. This has been invoked for the relationship between fig trees *Ficus* and their pollinating fig wasps. The reasoning is as follows. Each *Ficus* species is argued to be specific to a single, coevolved wasp species, as a route to exclude nonpollinating wasps (cheaters); and small variation among populations of a single *Ficus* species in their relationships with pollinating wasps is predicted to lead to reproductive isolation and speciation into new *Ficus*/fig wasp species pairs. The pairwise coevolution and cospeciation of the fig trees and their pollinators have been cited to explain the very high species diversity of the genus *Ficus*, comprising ~750 species worldwide in the tropics and subtropics.

Molecular analyses have revealed that this traditional interpretation of the *Ficus*/wasp relationship is wrong. The great advantage of molecular methods is that the divergence of two partners can be compared in the same units, e.g., nucleotide substitutions. Cospeciation is evident as, first, the same order of speciation in phylogenies of the two partners

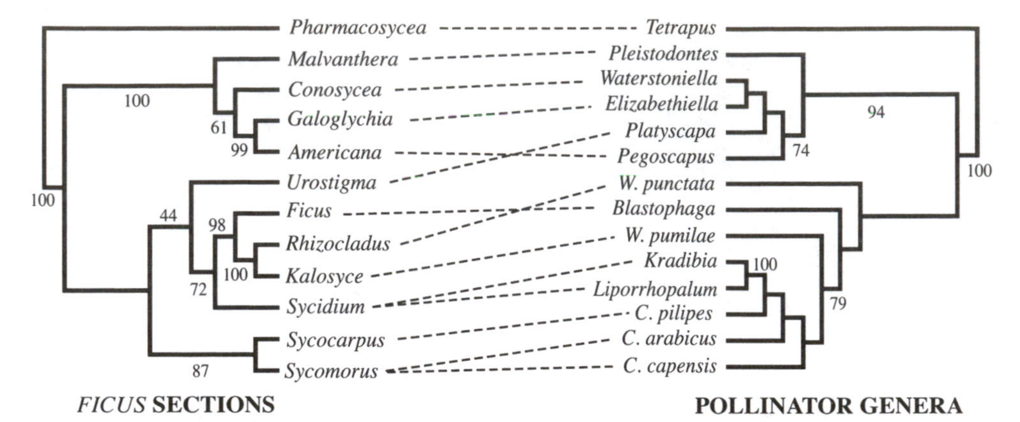

Figure 5-4 Comparison of the phylogenies of *Ficus* and their associated genera of pollinating wasps (including three species of *Ceratosolen* (C.) and two *Wiebesia* (W.) species). Bootstrap values (>40%) are shown on branches. [Reproduced from figure 1 of Machado et al. (2005). Copyright (2005). National Academy of Sciences, U.S.A.]

(giving congruent phylogenies), and, second, the same time of speciation (giving the same relative ratio of time between speciation events in the phylogenies). When these techniques were applied to the *Ficus*/fig wasp association, they revealed a much looser coupling between the plant and pollinator taxa than envisaged previously. Some *Ficus* species are pollinated by more than one wasp species, genetically indistinguishable wasps (presumably of a single species) pollinate multiple *Ficus* species, and regular host switches among the pollinators are evident from the molecular phylogenies (figure 5-4). It is not understood fully how these coevolutionary interactions without tight coupling contribute to speciation patterns of *Ficus* (Machado et al. 2005).

The *Ficus*/fig wasp associations are one of a variety of relationships in which the coevolutionary interactions are diffuse (Stanton 2003; Bascompte and Jordano 2007), meaning that the reciprocal interactions occur between groups of similar species with similar, general benefits. A mechanism for diversifying coevolution in such relationships has been proposed by the Geographical Mosaic Theory of Coevolution developed by Thompson (2005). Thompson's ideas are relevant to symbioses in which the impact of symbionts on host fitness varies across host populations. For example, certain endophytic fungi promote plant tolerance of extreme abiotic conditions, such as drought, high salinity, and high temperature, and these associations are particularly advantageous to plants in the various stressful arid, maritime, and geothermal habitats. From the landscape perspective, the associations are subject to a selection mosaic, with greater selection for reciprocal evolutionary change of the plants and fungi to enhance fitness in the extreme habitats. Thompson (2005) calls these habitats coevolutionary hotspots. If a coevolutionary hotspot remains fixed for extended periods and there is little gene flow between the hotspot and other locations where selection for coevolutionary change is small, reproductive isolation and species formation can arise. Even where their geographical distribution changes through time and coevolutionary changes do not lead to speciation, hotspots can contribute to the total genetic diversity of the interacting species. One important implication is that symbiotic partners do not have to be coevolving throughout their range for coevolution to be important. Selection mosaics are potentially important in the diversification of symbiotic organisms, in relation to traits that influence ecologically important traits of their partners (e.g., pathogen resistance, tolerance of abiotic conditions), but this topic has yet to be explored empirically in any detail.

Tight cospeciation is predicted in obligately vertically transmitted symbioses. For various mycetocyte symbioses between insects and their vertically transmitted bacteria (introduced in chapter 2, section 2.3.3 and table 2-1), the phylogenies of host and symbiont are perfectly congruent; this applies for example to the symbioses between aphids and *Buchnera*, ants and *Blochmannia*, and cockroaches and *Blattabacterium* (Moran et al. 1993; Sauer et al. 2000; Lo et al. 2003). The association by descent for these symbioses can be attributed to the mode of symbiont transmission. Because these symbionts are all transmitted strictly from mother to offspring, the symbionts of necessity diverge with their host lineages.

What is the evolutionary significance of cospeciation in vertically transmitted insect-microbial symbioses? At present, we do not know. The implications are immense if selection for host-symbiont interactions varies among host species. As a hypothetical example, host speciation events linked to changes in diet may lead to selection for changes in nutritional interactions between host and symbionts. This could result in the coevolution of exquisitely coadapted host-symbiont pairs with the potential to promote further diversification. Alternatively, host speciation may involve little or no functional change in the host-symbiont interaction, and coadapted symbioses may even be under strong stabilizing selection for no changes in the interactions. Under these circumstances, cospeciation of hosts and symbionts would have no impact on host phenotype or diversification rates. Resolving this uncertainty requires advances in our understanding of the specificity of vertically transmitted symbioses, as is discussed in chapter 4 (section 4.3).

5.2.3 Wolbachia: *A Special Case of Symbiosis-Induced Speciation*

One microorganism that has been suggested repeatedly to play a role in speciation of its hosts is the bacterium *Wolbachia* in arthropods, especially insects. The putative mechanism is very simple: that *Wolbachia* generates reproductive isolation as a by-product of its manipulation of host reproductive systems to promote its own transmission.

The principal focus of research on the evolutionary significance of *Wolbachia* has been cytoplasmic incompatibility (CI), where crosses between infected males and uninfected females are nonviable (see chapter 3, table 3-2). The current consensus is that insect speciation mediated solely by CI is unlikely to be important (Hurst and Schilthuizen 1998) because it poses an inadequate barrier to gene flow. For example, in the models of Turelli (1994), CI yields unstable equilibria comprising host populations with an intermediate prevalence of *Wolbachia* and not two reproductively isolated host populations, one infected and the other uninfected.

There is, however, evidence that *Wolbachia* can contribute to reproductive isolation between populations of their hosts by selecting for hosts to discriminate against matings which will yield inviable offspring. Jaenike et al. (2006) have explored the mating behavior of two *Drosophila* species, *D. subquinaria* and *D. recens*, with overlapping ranges in Canada. *D. recens* is infected with *Wolbachia*, which causes CI in heterospecific matings with *D. subquinaria*, leading to nonviable matings between male *D. recens*, and female *D. subquinaria*. Females of *D. subquinaria* from sites where *D. recens* also occur (sympatric populations) are very choosy, refusing to mate with male *D. recens*, but approximately 25% of females from sites where *D. recens* is absent (allopatric populations) mate with *D. recens*. Parallel molecular analyses indicate that the sympatric and allopatric populations of *D. subquinaria* are not genetically differentiatied, and this means that the behavioral difference between them can reasonably be attributed to selection on the sympatric populations to avoid *Wolbachia*-mediated failure to produce offspring. The really interesting consequence of the selection on sympatric *D. subquinaria* to be choosy is that their choosiness appears to be general. These flies discriminate against allopatric *D. subquinaria* males (table 5-1). In other words, *Wolbachia* infection of one insect species can contribute to reproductive isolation within a second, uninfected species.

Wolbachia occurs in many insect species; estimates vary, but ~20% of all 1–10 million insect species are likely to be infected at any one time. Many authorities consider the *Wolbachia* to be of little general significance as an agent of speciation in insects, but the study of Jaenike et al. (2006) on *Drosophila* species demonstrates that this bacterium can

Table 5-1
Impact of Selection to Avoid *Wolbachia*-Mediated Nonviable Matings on Mate Choice by Female *Drosophila subquinaria*

	Female D. subquinaria*: % mated*	
Male Drosophila	*Sympatric*	*Allopatric*
D. recens	0	25
Sympatric *D. subquinaria*	58	70[c], 57[i]
Allopatric *D. subquinaria*[1]	3[c], 21[i]	96[c], 83[i] (see footnote-1)

Source: Jaenike et al. (2006).

Female *D. subquinaria* sympatric with *D. recens*, which bear *Wolbachia*, are more choosy than females from allopatric populations (i.e., sites where *D. recens* is absent). The two allopatric populations of *D. subquinaria* analyzed were coastal ([c]) and inland ([i]).

[1]Matings between allopatric females and allopatric males refer to matings between flies of the same populations. More than 70% of females from the coastal and inland allopatric populations mated with males from the other allopatric population.

potentially have impacts of species diversity that are not immediately obvious and that extend beyond the taxa that it infects.

5.3 Symbiosis and the Structure of Ecological Communities

Mutually beneficial symbioses are not alternatives to antagonistic relationships because, as explained in chapter 1 (section 1.2), symbioses evolve and persist in the context of antagonistic interactions. This is particularly evident in relation to the impact of symbioses on the structure of ecological communities. For example, where two competing organisms differ in their responsiveness to symbiosis, their competitiveness will depend on whether the symbiotic partner is present. The consequences of such interactions can be profound.

Scleractinian corals in many shallow-water habitats are in competition with macroalgae. The corals are dominant where herbivorous fish are abundant and the waters are nutrient-poor (corals are more tolerant than macroalgae of low nutrient conditions). Where the waters are relatively nutrient-rich, the macroalgae are at a competitive advantage and tend to overgrow the corals. The competitive dominance of macroalgae in some habitats can be reversed by a symbiosis between corals and small crabs. At one high-latitude site in the United States, colonies of the coral *Oculina arbuscula* bear the crab *Mithrax forceps*. The crab gains protection from predators, and it promotes coral growth and survival by feeding on the competing macroalgae (Stachowicz and Hay 1999). One consequence of the reduced macroalgal biomass in the immediate environs of the coral is that the coral becomes accessible as a habitat for hundreds of other animal species. The coral-crab symbiosis transforms the local conditions, forming the basis for a dramatic change in the wider community.

The impact of symbiosis on competition is more subtle where the competing organisms are all symbiotic but vary in the benefit they derive from the symbiosis. This situation arises for plants in association with mycorrhizal fungi. Plant communities in low-nutrient soils are shaped largely by competition for nutrients. Mycorrhizal fungi promote plant access to mineral nutrients, particularly phosphorus, but there is considerable interspecific variation in plant response to the mycorrhizal symbiosis for nutrient acquisition. In prairie grassland communities, the dominant grasses are more responsive than the subordinate dicot species to mycorrhizal colonization, and the presence of mycorrhizal fungi results in enhanced dominance by the grasses and reduced plant diversity (Hartnett and Wilson 1999). In British meadow communities, the subordinate dicots are more mycorrhiza dependent than the grasses, and the impact of mycorrhizal fungi is reversed: the symbiosis promotes nutrient acquisition by the subordinate species, so reducing competitive dominance and increasing plant diversity (Grime et al. 1987; Gange et al. 1990). Plant diversity is particularly enhanced in soils bearing multiple mycorrhizal fungal species, probably because the species-rich mycorrhizal community exploits soil phosphorus efficiently and gives each plant species greater partner choice (van der Heijden et al. 1998).

Symbioses involving protection of an organism from predators can also have substantial impacts on ecological communities because they affect the food choice of the predators, with potentially substantial consequences for the wider ecological community. Here I consider terrestrial protection systems involving ants and endophytic fungi.

Some ant species that protect hemipteran insects from predators also consume other arthropods, thereby affecting the arthropod community on plants. For example, Mooney (2006) has demonstrated that the ant *Formica podzolica* both tends two aphid species and removes other arthropods from the canopy of pine trees *Pinus ponderosa*, resulting in increased abundance of the tended aphid species at the expense of other arthropods; and Vrieling et al. (1991) have shown that the ragwort plant *Senecia jacobaea* is protected from the defoliating larvae of the cinnabar moth *Tyria jacobaeae* when it bears populations of the aphid *Aphis jacobaea* tended by *Lasius* ants. In each of these examples, however, the community consequences of the hemipteran-ant association are complex.

For the pine system studied by Mooney (2006), the source of complexity is birds, including chickadees, nuthatches, and warblers, which also consume arthropods living in the pine canopy. The ants avoid foraging on branches that can be reached by the birds, leaving aphids on these branches vulnerable to attack by both birds and arthropod predators. As a consequence, the ants promote the abundance of tended aphid species only on branches where the birds cannot forage.

For the ragwort system, the complexity arises from the interaction between the ant-aphid symbiosis and the plant pyrrolixidine alkaloids which provide protection against most herbivores other than the defoliator *Tyria jacobaeae*. (As mentioned above, the ants tending the aphids protect the plant from *T. jacobaeae*.) *Aphis jacobaea* can detoxify the alkaloids (Hartmann 1999) but occurs preferentially on plants of low alkaloid content (Vrieling et al. 1991). Because aphid infestation reduces seed set of the plant, the aphids select for high-alkaloid plants; but because *T. jacobaeae* abolishes seed production, aphid-mediated ant protection selects for low-alkaloid plants. These contrary effects play out as selection for low-alkaloid plants when *T. jacobaeae* is abundant and high-alkaloid plants when *T. jacobaeae* is uncommon. In this way, the co-existence of plant genotypes with different alkaloid profiles can be attributed to spatiotemporal variation in the selective advantage of the aphid-ant symbiosis.

The protective endophytic fungi in grasses can also have community-level impacts, as has been explored in a long-term study of Clay and colleagues. Here, the impact of the endophyte-grass symbiosis on the wider plant community is mediated through the food choice of predators. Endophytic fungi in various grasses contain toxic alkaloids that protect the grass host from herbivory. Clay and Holah (1999) established replicated plots containing ten species of grasses and forbs, to which they introduced the grass *Lolium* (=*Festuca*) *arundinacea* either infected with the fungus *Neotyphodium coenophialum* or uninfected. Over the subsequent four years, the number of plant species recorded in the plots varied, but became progressively more depressed in plots with the endophyte-infected *L. arundinacea* than with the uninfected *L. arundinacea* (figure 5-5a). Parallel long-term analyses of the impacts of this grass on vegetation at two further sites confirmed the negative impact of the endophyte-grass symbiosis on plant diversity and, further, revealed that the endophyte-infected *L. arundinacea* caused the suppression of invasion by tree seedlings and stabilization of the grassland community (Rudgers et al. 2007). At two different sites, the abundance and height of tree saplings were depressed in plots enriched with the endophyte-infected *L. arundinacea* relative to those with uninfected *L. arundinacea*. In effect, the symbiosis delayed successional change of grassland to woodland. These effects on plant diversity can be attributed principally to the fact that herbivores respond to the toxicity of the

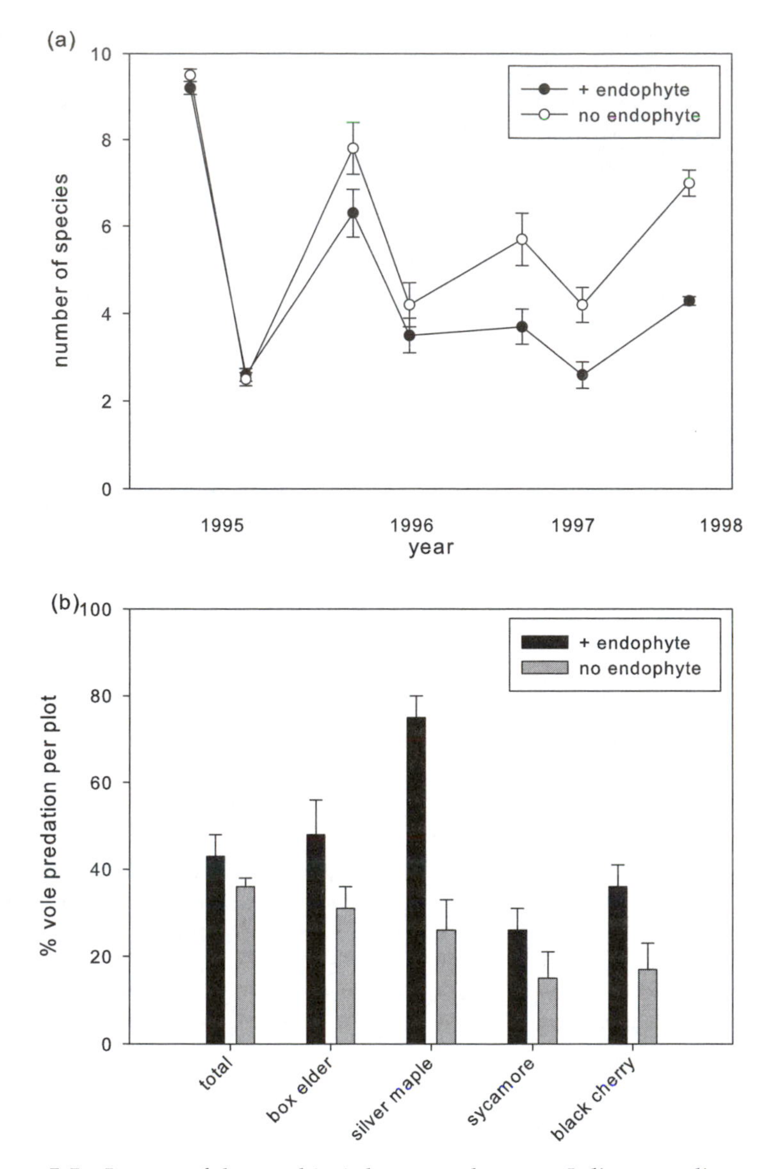

Figure 5-5 Impact of the symbiosis between the grass *Lolium arundinacea* and endophytic fungus *Neotyphodium coenophialum* on the plant community. (a) Number of plant species over four years after introduction of *L. arundinacea* to a mixed grass/forb community. (b) Predation of tree seedlings by voles in plots containing *L. arundinacea*. [Reproduced from *Science*, figure 2A of Clay and Holah (1999) and figure 3A of Rudgers et al. (2007). Reprinted with permission from AAAS.]

endophyte-infected *L. arundinacea* by increased consumption of other plants (figure 5-5b).

The impact of endophyte-infected *L. arundinacea* on the plant community is of particular interest because this grass species is an exotic and highly invasive in natural grasslands of the Midwest of the United States, where the experiments of Clay and colleagues were conducted. The effect of this symbiosis on native vegetation is compounded by the response of herbivores, which generally prefer to eat native plants. In the study of Rudgers et al. (2007), the one tree species that was not suppressed in plots with endophyte-infected *L. arundinacea* was an introduced species, the white mulberry. In this way, the endophytic fungal symbiosis can alter the species composition of the incipient woodland to favor non-native species. Symbiosis has been implicated repeatedly in the establishment and invasiveness of introduced plant species, and I return to consider this issue in the next section. In the present context, the importance of these studies is that they demonstrate how mutually beneficial symbioses can have a very substantial impact on the structure of ecological communities as a direct result of the nutritional or protective services provided.

5.4 The Success of Symbioses in the Anthropocene

5.4.1 The Anthropocene

These are special times, when human activities are so pervasive that they dominate global processes on the planet. The five key anthropogenic factors are large-scale extinction, biotic homogenization through invasive species, habitat fragmentation, climate change, and modification of biogeochemical cycles. Because the impacts are so large, some authorities are treating current conditions as a geological era, the Anthropocene, equivalent to the Holocene (the last 10,000 years) or the Pleistocene. The Anthropocene is variously considered to have started at the Industrial Revolution (ca. 1800) or ~5000 years ago (Crutzen 2002; Ruddiman 2005). The Anthropocene is an excellent descriptor for the twenty-first century but its duration and geological significance will depend on the reversibility of current anthropogenic impacts and persistence of very large human populations.

How successful are symbioses in the Anthropocene? As explained in the Introduction to this chapter, it can easily be argued that symbioses are especially vulnerable to at least one anthropogenic factor, extinction: the local or global extinction of one organism in symbiosis is expected to put its partners at risk and, because many symbioses have important ecological roles, their disruption will have substantial effects on wider ecosystem processes. As a result, the loss of one partner in a symbiosis can potentially trigger a cascade of linked extinctions through the ecological community (Gilbert 1980; Bond 1994). Despite its appeal, this argument may not be universally valid. We know that some symbioses are evolutionarily remarkably persistent, including through major climatic changes and, as discussed in chapter 2 (section 2.4.2), the crucial comparisons between the extinction rates of symbiotic and related nonsymbiotic taxa have not been conducted systematically (Sachs and Simms 2006). Furthermore, there is accumulating evidence that some symbioses are contributing to anthropogenic extinctions by promoting biological invasions that are deleterious for native taxa, as is discussed below.

High-quality field data are crucial to assess the global impacts of human activities on symbioses and these are available for the three types of relationship that I consider: the seed dispersal mutualisms, corals, and plant-fungal symbioses. Together, they reveal that anthropogenic factors affect the success of symbioses in many different ways and with a range of consequences.

5.4.2 *Vulnerability of Symbioses to Partner Decline*

The decline or extinction of one symbiotic organism is predicted to lead to the extinction of its partners where the association is highly specific and obligate. As considered in chapter 2 (section 2.4.2), this argument may not be universally valid because some organisms in apparently obligate associations have, over evolutionary timescales, switched partners. Even so, there is a general expectation that associations of low specificity are robust to the loss of a symbiotic partner. Indeed, as discussed in chapter 4 (section 4.3), low partner availability is seen as a likely selection pressure for low specificity.

Let us now turn to anthropogenic effects, especially those, such as habitat fragmentation and invasive species, which tend to reduce biodiversity. Overall, these effects are more likely to be tolerated by relationships of low specificity; an organism with low specificity can respond to the local extinction of one partner by switching to other partners. A particularly good system to test this argument is provided by a nonsymbiotic mutualism, seed dispersal, which characteristically has very low specificity. Multiple animal taxa disperse the seeds of any one plant species, and each plant species attracts multiple species of dispersers, resulting in a "loose interdependence" between guilds of plants and dispersers (Herrera 2002). The underlying reason is that it is in the selective interests of neither partner to be specific: associating with multiple species of the partner enhances food sources for the animal and dispersal opportunities for the plant.

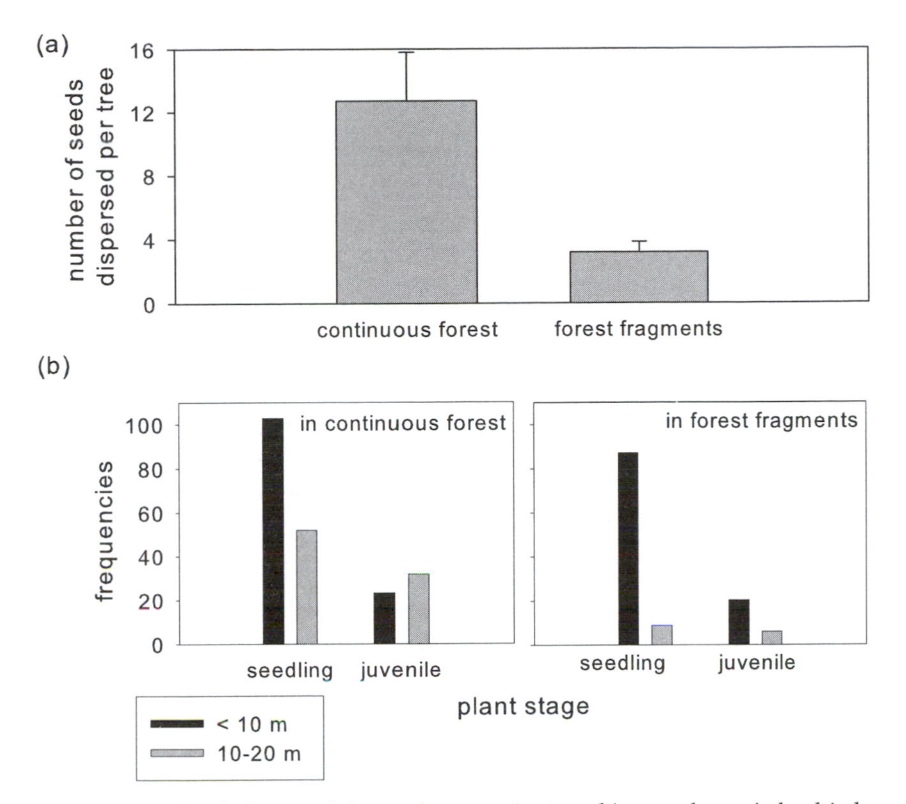

Figure 5-6 Seed dispersal from the tree *Leptonychia usambarensis* by birds in continuous and fragmented forests of Tanzania. (a) Numbers of seeds dispersed. (b) Dispersal distance and survivorship to juvenile stage. [Redrawn from figures 1 and 3 of Cordeiro and Howe (2003). Copyright (2003). National Academy of Sciences, U.S.A.]

The empirical data do not support the expectation that seed dispersal mutualisms are robust to partner decline, as is illustrated by two elegant field studies on bird-mediated and ant-mediated dispersal.

Cordeiro and Howe (2003) focused on a single midstory tree species *Leptonychia usambarensis,* which is an endemic of the Eastern Arc and Coastal Forests of Kenya-Tanzania and is dispersed by a variety of bird species. Parts of the forest have become fragmented through increased human activity, resulting in reduced bird diversity. Although the seed dispersal mutualism is nonspecific, seed dispersal is reduced by 75% in the forest fragments (figure 5-6a), resulting in significantly fewer seedling and juvenile plants at a distance from the trees in the fragmented sites than in the continuous forest sites (figure 5-6b). Overall, more than 90% of all the seedlings in the fragmented sites were scored under the crown of the parent tree (<10 m), compared to 64% for continuous forest sites, indicative of dispersal limitation in the fragmented sites. Furthermore, the low total frequency of juvenile trees in fragmented sites (figure 5-6b) suggests that recruitment of *L. usambarensis* is impaired by the impact of habitat fragmentation on bird-mediated seed dispersal. This detailed analysis of a mutualism in a single species demonstrates that the nonspecific seed dispersal mutualisms are not necessarily robust to reduced partner diversity. There is a very real concern that these processes are promoting the local extinction of multiple plant and animal species in loosely linked dispersal mutualisms (Brooks et al. 2002; Bascompte and Jordano 2007).

Seed dispersal mutualisms are also vulnerable to invasive species. The Argentine ant *Linepithema humile* was introduced to South Africa in the early 1900s and now has substantial populations in the fynbos vegetation. The seeds of about one-third of plant species in the fynbos are dispersed by native ants, which harvest the seeds, consume the elaiosome (food-body attached to each seed), and discard the viable seed underground. The Argentine ant can potentially out-compete the native ants and, thereby, disrupt seed dispersal and alter the plant community composition of the fynbos (Bond 1994). Field surveys by Christian (2001) revealed that the populations of ant genera dispersing relatively large seeds (>1 g) are depressed at sites colonized by the Argentine ant. Subsequent experiments demonstrated that dispersal of the large seeds was selectively reduced in the presence of Argentine ants (table 5-2a); most of the seeds that were not dispersed were eaten by rodents. Finally, the recruitment of the large-seeded plants was selectively depressed at sites colonized by the Argentine ant (table 5-2b), indicating that the Argentine ant can trigger a change in the community composition through its impacts on the seed disperser community.

The conclusions from these studies of seed dispersal mutualisms are likely to be relevant to many symbioses. It appears that, although organisms with low specificity are tolerant of the variable availability of individual partner species (chapter 4, section 4.3), they are not necessarily robust to the scale of biodiversity loss associated with the anthropogenic effects of habitat loss and invasive species. Extinction risk is not particular to highly specific associations, and these latter associations should not be the exclusive focus of conservation effort.

5.4.3 *Susceptibility to Climate Change*

In the popular and environmental press, coral reefs are routinely cited as one of the first casualties of global climate change, the "miner's canary" whose demise predicts a greater future catastrophe. The symbiosis between the corals and their intracellular dinoflagellate algae

TABLE 5-2
Impact of the Argentine Ant on the Plant Community in the Fynbos Vegetation

Plant species	*Sites*	
	With Argentine ant	*Without Argentine ant*
(a) Fate of experimental seeds	*% seeds dispersed*	
Large-seeded plant[1]	25	50
Small-seeded plant[2]	60	60
(b) Seedling recruitment[3]	*No. seedlings per adult plant* m^{-2}	
Large-seeded plants	1.14	7.46
Small-seeded plants	5.92	4.26

Source: Figures 4 and 5 of Christian (2001).
[1]*Leucopsermum truncatulum.*
[2]*Spatalla racemosa.*
[3]Combined data from three plant species with each of large and small seeds.

Symbiodinium is central to the corals' apparent susceptibility to climate change. Specifically, the symbiosis collapses in response to environmental change, especially elevated sea water temperatures. However, it should be recognized that the impact of bleaching on corals is compounded by other anthropogenic factors operating at scales ranging from local (e.g., industrial pollutants) to regional (e.g., increased coastal sedimentation linked to changes in land use such as deforestation) and global, particularly the predicted impact of acidification caused by increased dissolved CO_2 on coral calcification rates (Hoegh-Guldberg et al. 2007). The scale of the threats to coral reefs appears to be immense, with some authorities predicting the collapse of the coral reefs of the Indian Ocean by 2020 (Sheppard 2003) and their worldwide collapse by 2050 (Hoegh-Guldberg 1999). But our focus here is exclusively on the response of the symbiosis to environmental change.

Corals bleach in response to environmental change. Formally, bleaching refers to the loss of color from the hosts of symbiotic algae. In other words, bleaching is an imprecise term founded on the appearance of the symbiosis to the human eye. It usually involves the partial to total loss of the algal population that, for corals, leaves the white calcareous skeleton visible under the transparent animal tissues; and it can also arise from the destruction of algal pigments, for example by a sudden increase in light intensity, with little or no change in the algal density. Bleached corals do not invariably die. The photosynthetic pigments or

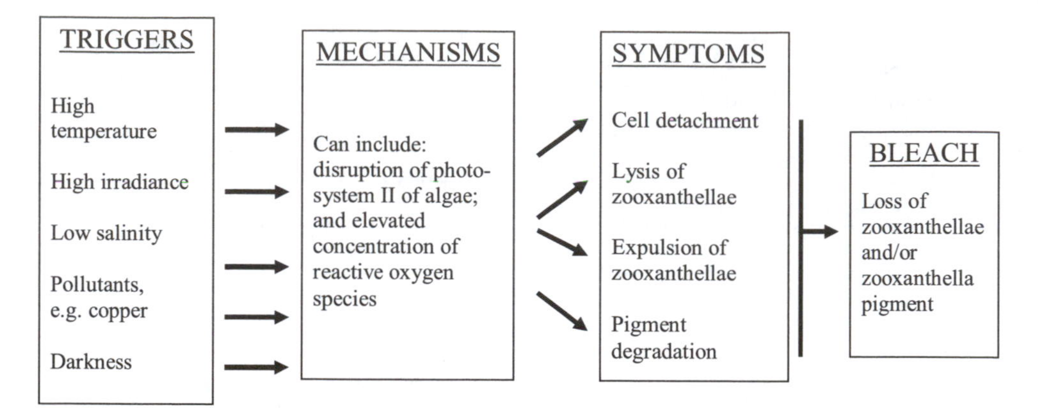

Figure 5-7 An explanatory framework to describe the causes of bleaching. [Reproduced from *Marine Pollution Bulletin*, figure 1 of Douglas (2003) with permission from Elsevier]

residual algal population in some corals recover over a period of weeks to months. Indeed, some apparently healthy corals in seasonal habitats undergo annual cycles of algal density that include periods when the corals appear very pale (Fitt et al. 2000). Nevertheless, bleached corals have dramatically reduced growth rates and are susceptible to disease and mechanical damage, and some bleaching events have resulted in mass mortality. Because corals are the foundation to coral reef ecosystems, coral mortality caused by bleaching is a first-order threat to biodiversity in tropical and subtropical seas (Hoegh-Guldberg et al. 2007).

To address the symbiotic dimension of coral bleaching, we need first to consider how bleaching occurs. This can be explored from three linked perspectives illustrated in figure 5-7. First, there are the environmental triggers. Elevated temperature, often associated with high irradiance, appears to be the most important trigger globally, but other triggers (e.g., low salinity, metal pollutants, pathogenic bacteria) can be locally important or act in combination with elevated temperature. A second way to describe the causes of bleaching is in terms of the symptoms. The symptoms include the loss of *Symbiodinium* cells, commonly by exocytosis from host cells. Entire host cells and their complement of *Symbiodinium* can also be sloughed off and lost. In some coral species, the expelled algal cells are alive (and often dividing actively) and in other species, they are dead, apparently dying by a process akin to programmed cell death. The third and crucial element is the cellular and molecular mechanisms by which the triggers are translated into these symptoms (figure 5-7). To date, most research on the mechanisms of bleaching has focused on the primary lesion caused by elevated temperature and irradiance. This lesion is in the *Symbiodinium*, specifically the thylakoid membranes of the chloroplast bearing photosystem II (PSII), which mediates the stripping of electrons from water to release molecular oxygen (Warner et al. 1999; Tchernov et al. 2004). One scenario is that excess excitation energy arising from temperature- or irradiance-induced inhibition of PSII generates reactive oxygen species (ROS), such as superoxides and oxygen radicals, which damage plastid thylakoids and other cell components. We are ignorant of subsequent molecular and subcellular events leading to the symptoms of bleaching. An important, unresolved issue is whether the mechanisms of bleaching triggered by different environmental factors have a core element in common. One scenario is that bleaching is mediated by host sanctions on algal cells that fail to provide nutrients, irrespective of the environmental trigger (see chapter 3, section 3.4.2). Bleaching may, additionally or alternatively, involve disruption of the capacity of the host immune system to manage the algal population (see chapter 4, section 4.5). At this stage, we cannot exclude the possibility that bleaching is an umbrella term for multiple, mechanistically distinct processes with superficially similar symptoms. Our limited understanding of bleaching mechanisms is important because it constrains our capacity to predict and manage the response of coral reefs to climate change.

What is the likely fate of corals and coral reefs? As considered earlier in this chapter (section 5.2.1), the scleractinian corals have been remarkably robust to dramatic ecological changes throughout their evolutionary history. In this context, their susceptibility to anthropogenic factors in recent decades is paradoxical. It can be argued that the current circumstances are uniquely detrimental to corals, with rapid climate change compounded by other effects, including disruption of both biotic interactions (e.g., by overfishing) and the physical habitat (eutrophication, sedimentation, etc). Alternatively, the corals may, over timescales greater than years to decades, be robust to anthropogenic factors through adaptation and experience-mediated changes that increase coral tolerance of anthropogenic impacts. There is persuasive evidence that at least some corals are more resistant to bleaching if they have experienced a previous bleaching event (Brown et al. 2000), and this may be mediated by either host-specific changes (Brown et al. 2002) or changes in the algal complement (Baker et al. 2004).

The apparent susceptibility of the coral-alga symbiosis to climate change should not be interpreted as evidence that symbioses are intrinsically more vulnerable than nonsymbiotic organisms to environmental perturbation. Indeed, the reverse is predicted for various organisms which gain thermal tolerance through symbiosis (e.g., Redman et al.

2002; Montllor et al. 2002). For symbiotic and nonsymbiotic taxa alike, we should expect many taxa to be prone to extinction and others to be successful in the Anthropocene.

5.4.4 Symbioses as Facilitators of Invasions

There is now overwhelming evidence that symbiosis can promote invasions by certain plants. Humans transport plants among the world's continents, either intentionally as for agriculture, forestry, and horticulture, or accidentally. A very small proportion of these plants establish self-perpetuating populations that spread from the sites of introduction, but these few invasive species are costly, both economically and in terms of their disruptive effects on native communities and biodiversity.

For a number of plant species, invasiveness has been linked to their interactions with microbial symbionts. As discussed in chapter 1 (section 1.4.4) and chapter 2 (section 2.3.2), the great majority of plants associate with mycorrhizal fungi, especially the arbuscular mycorrhizal (AM) and ectomycorrhizal (ECM) fungi. Generally, associations between introduced plants and native fungi are much more likely for AMs than ECMs. AM fungi were traditionally believed to be nonspecific, with every fungal isolate able to associate with any AM-compatible plant. Although many AM fungi are not universally promiscuous (see chapter 4, section 4.3), it appears that most soils bear sufficient generalist AM-fungal taxa to support introduced plants. The multitude of exotics in suburban gardens of the developed world is made possible by the compatibility of the native AM fungi and introduced plant species with no shared evolutionary history. In contrast, some plant species dependent on ECM fungi have failed to establish in novel habitats. The introduction of pines (*Pinus*) to the southern hemisphere has been hampered by the paucity of native ECM fungi, including their apparent absence from South Africa. For *Pinus* seedlings to grow in these soils, it has proved necessary to introduce fungal propagules. Various commercial companies market mycorrhial fungi for forestry and other plant production markets.

Richardson et al. (2000) report that most invasive plants associate with AM fungi. Further, they conclude that, although AM associations permit establishment in a novel habitat, attributes other than their mycorrhizal status are generally responsible for invasiveness. There are a few exceptions, however, where AM fungi appear to act as determinants of plant invasiveness. The Asian knapweed *Centaurea maculosa* is a noxious weed in natural prairie grassland habitats of the United States. Marler et al. (1999) provide evidence that *C. maculosa* interferes with the mycorrhizal networks on which co-occurring native grasses, including *Festuca idahoensis,* depend. In experimental plots, both *C. maculosa* and *F. idahoensis* formed mycorrhizas with AM fungal inoculum from prairie soils, although the growth of neither was dependent on the association. When the two plant species were grown together, their performance depended on whether the mycorrhizal fungus was present. Without the fungus, plant performance was unaffected by the other plant species and, with the fungus, the alien species *C. maculosa* performed better at the expense of the native species *F. idahoensis.* In other words, the deleterious impact of the introduced plant on the native plant could be attributed to its disruption of the indigenous mycorrhizal network. The underlying processes are not resolved fully (Zabinski et al. 2002; Callaway et al. 2004), and it remains to be established whether *C. maculosa* is acting as a cheater of mycorrhizas (see chapter 3, section 3.3) or causes relatively nonspecific community-level changes in soil organisms, including the native mycorrhizal network.

Forestry practice has probably contributed to the invasiveness of some ECM-dependent trees. As considered above, commercial inocula of mycorrhizal fungi are used extensively for exotic plantations, for example of the Monterey pine *Pinus radiata* (native of California and Mexico) in Australia and of *Eucalyptus* (from Australia) in the Mediterranean and United States. These trees are becoming invasive into native habitats close to the plantations. Importantly, their roots are colonized principally by ECM fungal taxa in commercial inocula or from the country of origin of the invasive plant (Diez 2005). There is a strong suspicion that these plant invasions are made possible by the establishment of the cointroduced ECM fungi in the native soils (Richardson et al. 2000; Schwartz et al. 2006).

The root symbionts are not the only mutualists that are implicated in plant invasions. The roles of shoot endophytes in promoting the persistence of invasive grasses and the invasive ants in the disruption of seed dispersal mutualisms have been considered earlier (see sections 5.3 and 5.4.2), and Richardson et al. (2000) review the contribution of pollination and seed dispersal mutualisms to various plant invasions. An important implication is that the traditional ecological explanations of invasions exclusively in terms of antagonistic interactions, especially competition, are inadequate and potentially misleading. The expectation is that invasive species are likely to be unrelated to any native taxa, and so lack competitors, and that low-diversity communities (which are believed widely to have relatively weak competitive interactions) are susceptible to invasion. The realization that mutually beneficial interactions, such as plant root symbioses, can promote establishment and invasiveness of introduced plants leads to very different expectations:

that the likelihood of compatible partners might be promoted, first, by functional similarity or close taxonomic relationship with native taxa and, second, by introduction to communities that are sufficiently species-rich to include compatible partners. We should expect the relative importance of beneficial and antagonistic interactions for the establishment and invasiveness of exotics to vary on a case-by-case basis, and the successful establishment and subsequent invasion of introduced species to be defined by interactions among multiple factors. A grasp of the evolutionary history and specificity of symbiotic organisms in different geographical regions and habitats, as discussed in chapters 2 and 4, is essential to obtain reliable predictions of the invasive potential of introduced species.

5.5 Harnessing Symbioses for Human Benefit

From the human perspective, one measure of the success of symbioses is our capacity to manage and manipulate them to our own advantage, based on a rational understanding of symbiotic interactions. Symbioses are being exploited in multiple contexts, including medicine, drug discovery, food production, pest control, habitat management, and remediation of degraded habitats (e.g., polluted soils). Here, I focus on the application of symbioses to promote the discovery of bioactive compounds, in novel insect pest management strategies and in human health. The applications in pest control and human health have the common feature that they seek to manipulate the symbiosis by modifying natural associations to obtain desired host traits (e.g., enhanced health, disease vector incompetence). In other words, symbiosis is used to obtain robust changes in host phenotype without genetic manipulation of the host. Some of the strategies I discuss are relatively well developed, and others have barely progressed beyond a good idea.

5.5.1 Symbiosis as a Source of Bioactive Compounds

Symbiotic organisms have great potential as a source of bioactive compounds with value as either pesticides or therapeutic agents, especially as novel antimicrobials and cancer drugs. Compounds with novel chemistries are in demand especially for antimicrobials because current antibiotics are founded on a very limited set of drug classes to which an increasing number of pathogens are resistant. Novel pest management strategies are also needed as a result of increasing pest resistance and concerns about the environmental and health impacts and safety of some broad-spectrum agents currently in use.

Microorganisms in symbiosis are predicted to include a high incidence of taxa that synthesize bioactive compounds. Let us consider why. Bioactive compounds are generally produced by microorganisms that need them for their own protection. These microorganisms tend to be slow-growing and to live in highly structured environments where they can, by chance, become locally abundant and highly competitive against faster-growing nontoxic taxa (Wiener 2000). For example, most known antibiotic-producing microorganisms have been isolated from soil, a highly structured environment made up of many separate pores among the soil particles, where each pore has distinct conditions and resources. Hosts are also highly structured environments where slow microbial growth rates are at an advantage, and we should expect some symbiotic microorganisms to also produce bioactive compounds.

The symbiotic habit offers additional advantages that facilitate the discovery and initial analysis of candidate bioactive compounds. Microorganisms reliably associated with particular species of plant or animal are readily accessible and, for vertically transmitted and other specific symbioses, related taxa with similar chemistries can be recovered from related hosts, assisting the study of multiple candidate compounds and their mode of synthesis. Where the metabolic pathways by which compounds of interest are synthesized are known, the relevant genes can then be transferred to more amenable organisms for detailed structure-function analysis, product modification, and, ultimately, commercial production. Research is, thereby, not constrained by difficulties in culturing the organism that produces it naturally or by the tendency for symbiotic microorganisms to cease production of the bioactive compound when isolated from the association. The genetic basis is established for the synthesis of several products by symbiotic organisms, including polyketides and patellamides (Piel et al. 2004; Donia et al. 2006). Metagenomic approaches, i.e., high-throughput sequencing of DNA from environmental samples containing multiple organisms, offer additional leverage for gene discovery and analysis in symbioses because they enable the identification of long biosynthetic pathways comprising tens of genes in complex microbial communities associated with animals or plants.

Several products derived from symbioses are in preclinical or clinical trials. They include a candidate antimalarial, manazamine produced by the actinomycete *Micromonospora* sp. isolated from the sponge *Acanthostrongylophora* [research of R. T. Hill et al., cited by Wijffels (2007)]; and the polyketide bryostatin, a candidate antitumor agent, produced by the bacterium *Endobugula sertula* in the marine bryozoan *Bugula neritina*. Another set of promising leads relate to the endophytic fungi of plants, which are being screened and developed as a source of novel pesticides (Strobel and Daisy 2003). For example, the fungus *Muscodor albus* originally isolated from the *Cinnamomum* tree in Central America produces a complex mixture of volatile organic compounds (including lipids, esters, alcohols, and ketones) that is active against a variety of pathogens. Mycofumigation with these volatiles is being used to protect fruit during storage and is under consideration for soil sterilization as an alternative to methyl bromide, an agricultural fumigant that is now banned in many countries (Strobel 2006).

5.5.2 Symbiotic Organisms in Novel Insect Pest Control Strategies

Symbiotic microorganisms of insects have great potential as novel strategies for insect pest management. Three broad approaches are available: exploitation of the cytoplasmic incompatibility (CI) induced by the bacterium *Wolbachia*, disruption of symbioses required by insect pests, and symbiont-mediated manipulation of insect traits.

Wolbachia-induced CI is expressed as sterile matings between infected males and uninfected females (see chapter 3, section 3.5.1 and table 3-2). If infected males are released to excess into an uninfected population, most females are predicted to mate with these males and produce inviable eggs. Proof-of-principle for this approach has been obtained by Zabalou et al. (2004), working with the Mediterranean fruit fly *Ceratitis capitata*. This species generally lacks *Wolbachia* infections, but can be infected experimentally with a *Wolbachia* from the related cherry fruit fly *Rhagoletis cerasi*, causing CI. In population cages with *Wolbachia*-infected and uninfected males at 50:1 ratio, the *C. capitata* population was suppressed by 99% in a single generation (figure 5-8). There is potential to apply this approach to populations of insect pests that naturally lack *Wolbachia*, including the olive fly *Bactrocera oleae*, and *Anopheles* mosquitoes. However, the ability of *Wolbachia* strains to colonize and persist in novel insect hosts cannot, at present, be predicted and the construction of novel combinations involves serendipity and hard work. The symbiotic problem is to identify the determinants of specificity (chapter 4, section 4.2), so that the appropriate *Wolbachia*-insect combinations can be selected rationally. This task is anticipated to become more tractable because the relative ease with which bacterial genomes can now be sequenced should facilitate the identification of *Wolbachia* genes contributing to host compatibility.

An indication of the likely value of CI-based control comes from its similarity to the sterile insect technique, which can eliminate local populations of pests within a generation by the release of saturating numbers of reproductively sterile males. The principal risk of CI-based control (but not the sterile insect technique) is the inadvertent release of

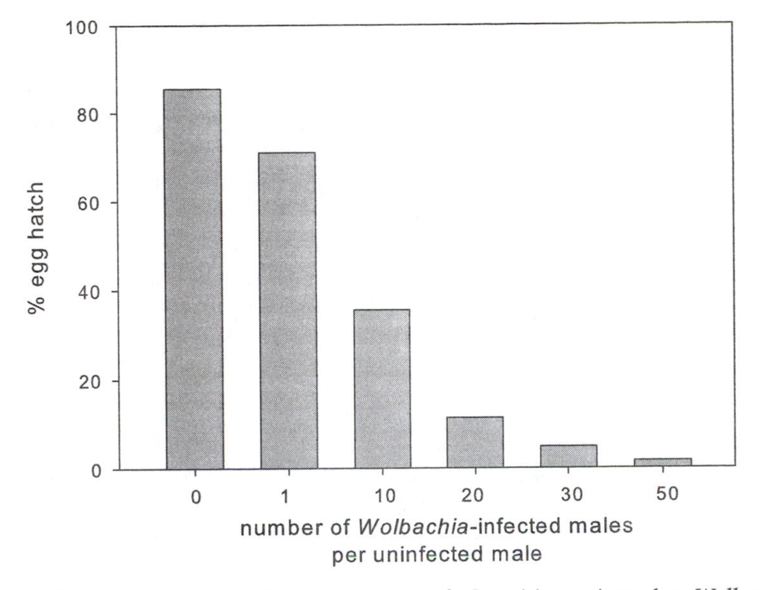

Figure 5-8 Suppression of populations of *Ceratitis capitata* by *Wolbachia*-mediated cytoplasmic incompatibility. In all populations the ratio of uninfected males to uninfected females was 1:1. [Redrawn from figure 2 of Zabalou et al. (2004). Copyright (2004). National Academy of Sciences, U.S.A.]

Wolbachia-infected females, which would be compatible with both infected and uninfected males, leading to the spread of *Wolbachia* through the wild population and collapse of the entire control strategy. CI-based control therefore depends on absolute confidence in the sexing and separation of male and female insects before the mass release of males.

Symbiosis disruption has great potential for insect pests bearing mycetocyte symbioses (chapter 2, section 2.3.3) because the microbial symbionts are required by the insect and obligately vertically transmitted via the ovary. As a result, an insect deprived of its symbionts cannot acquire an alternative supply from the environment or other insects. Mycetocyte symbionts can be eliminated by antibiotic treatment and the resultant insects grow poorly and are reproductively sterile. As a result, symbiosis disruption is suitable for situations requiring the suppression of insect pest populations over periods of days to weeks but not for immediate insect death. The commercial use of antibiotics is unacceptable on both economic and public health grounds, and the challenge is to identify alternative routes to disrupt these symbioses. Two complementary approaches are available: to identify specific targets, and to screen for symbiosis-active compounds. From the whole-organism perspective, the ideal targets are symbiont transmission to the eggs in the ovary and nutrient exchange between insect and symbiont. Identification of targets at the molecular level is increasingly becoming feasible because the availability of genomic data for various symbionts and their insect hosts (e.g., Zientz et al. 2004; Dale and Moran 2006) provides the basis for the molecular dissection of symbiosis function.

The search for compounds active against mycetocyte symbioses has barely started. A likely source is the plant kingdom, since plants with an effective defense against phytophagous insects with mycetocyte symbionts would be at a selective advantage. In one study, certain labiate plants depressed the performance of aphids containing the usual complement of the mycetocyte bacteria *Buchnera*, but not antibiotic-treated aphids lacking *Buchnera*; and this poor growth was linked to reduced nutrient release from the bacteria (Wilkinson et al. 2001). It is an open question whether the class(es) of compounds active against symbioses in phytophagous insects will have activity against other mycetocyte symbioses, including those in blood-feeding insects. Although the various symbioses have evolved independently and involve phylogenetically different microorganisms, there may be overlap in the underlying molecular mechanisms through convergence, leading to the availability of compounds active against a range of insect pests of agricultural, medical and veterinary importance. The involvement of a conserved signaling pathway in multiple plant symbioses and similar cellular processes in the persistence of beneficial and pathogenic associations (chapter 4, sections 4.2.4 and 4.5.3) illustrates how common processes can underpin multiple different associations.

The final strategy of symbiont-mediated manipulation of insect traits is not to eradicate the pests but to eliminate their harmful effects, e.g., by compromising their competence as vectors of disease or by restricting their host range so that they are unable to utilize the target hosts (humans, crops, etc.). This strategy requires microorganisms that confer the traits of host range restriction or vector incompetence, together with routes for their rapid dissemination through the pest population. To date, most progress has been made by introducing traits to symbiotic microorganisms by genetic manipulation. This approach is termed paratransgenesis, defined formally as the modification of the insect phenotype by genetic transformation of its associated microorganism(s) (Ashburner et al. 1998), and it is particularly suitable for microbial symbionts which can be cultured and transformed readily.

The efficacy of paratransgenesis has been demonstrated experimentally for one system, the blood-feeding reduviid bug *Rhodnius prolixus*, which is the vector of *Trypanosoma cruzi*, the agent of Chagas disease. This trypanosome is located in the insect gut and, while taking a blood meal, the insect deposits fecal pellets containing live trypanosomes and

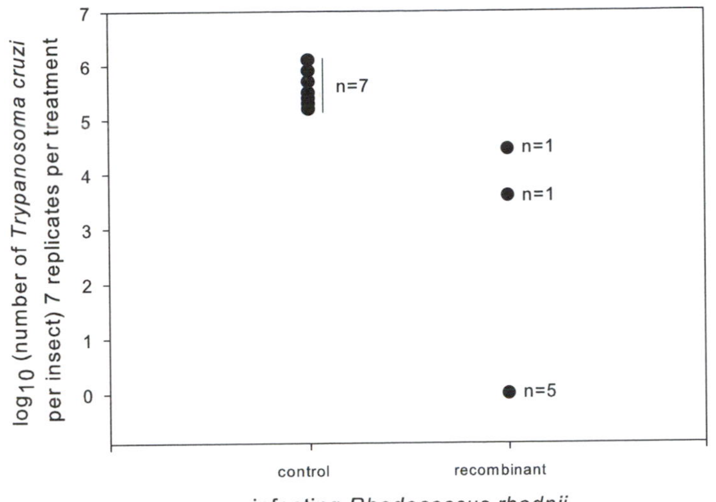

Figure 5-9 Number of *Trypanosoma cruzi* in *Rhodnius prolixus* infected with recombinant *Rhodococcus rhodnii* bearing the cecropin A gene and the native (control) *R. rhodnii*. [Redrawn from figure 5 of Durvasula et al. (1997). Copyright (1997). National Academy of Sciences, U.S.A.]

other gut microorganisms onto the skin of the host. Durvasula et al. (1997) have transformed a dominant member of the gut microbiota of the insect, the streptomycete *Rhodococcus rhodnii*, with the gene for cecropin A, a peptide which lyses *T. cruzi*. When the transgenic bacteria were introduced to the insects, transmission of *T. cruzi* was reduced by more than 1000-fold in most insects tested (figure 5-9). Dissemination of the transgenic bacteria is achieved through the insect's feeding habits: the insects regularly eat fecal pellets, such that the transgenic bacteria, introduced to a small number of insects, are readily distributed throughout the local pest population. Durvasula et al. (1997) have exploited this habit by constructing artificial feces inoculated with the transgenic *R. rhodnii* suitable for spraying into habitations infested with the insects. The product is known as CRUZIGARD and field trials of its efficacy in controling Chagas disease are being explored.

The regulatory requirements and public concerns about genetic modification (GM) technology may restrict the opportunity for pest manipulation by paratransgenesis (Hoy 2000). In particular, the risk of transgene escape represents a substantial technical challenge, given the high proliferation and dispersal capacity of microorganisms and the incidence of horizontal gene transfer among unrelated bacteria, especially in invertebrate gut environments (Thimm et al. 2001).

An alternative approach is to harness naturally occurring microorganisms to manage insect pests. This topic has attracted research attention in relation to few pest taxa, notably mosquitoes and aphids. There is some evidence that the gut microbiota in *Anopheles* mosquitoes has a protective role against the malaria parasite *Plasmodium*. When the mosquitoes were fed with the antibiotic gentamycin, they were infected more readily with the malaria parasite *Plasmodium*; and *Plasmodium* infections were also depressed when the mosquitoes were co-fed with *Plasmodium* and Gram-negative bacteria in comparison to *Plasmodium* only (Pumpuni et al. 1993; Beier et al. 1994). These experiments raise the possibility that interventions to manipulate the gut microbiota could suppress vector competence of mosquitoes.

Interest in harnessing naturally occurring microorganisms to control aphids comes from the growing evidence that some aphid traits are shaped by whether they bear specific bacteria, known generically as secondary symbionts. These bacteria are not required by the insect and occur at variable frequencies in natural populations. (They are distinct from the bacterium *Buchnera* required by aphids, discussed above.) The secondary symbionts variously promote resistance to biological control agents, including parasitic wasps and pathogenic fungi, reduce the capacity of aphids to use particular plants, and promote aphid tolerance of high temperatures (reviewed in Dale and Moran 2006). This raises the opportunity, in principle, to reduce aphid pest status by manipulation of the secondary symbiont complement. For example, the biological control of aphids by parasitic wasps or pathogenic fungi would be enhanced by strategies which depress the incidence of secondary symbionts conferring resistance to these control agents, and crop damage would be reduced by promotion of secondary symbionts that exclude the crop from the aphid plant range. The application of these approaches requires strategies to manipulate the bacterial prevalence in the insect pest populations.

Many insects bear facultative microorganisms in their gut or tissues, raising the possibility that the traits of other pest taxa may be influenced by their microbiota. For example, there is preliminary evidence that the secondary symbiont *Sodalis glossinidius* in the tsetse fly *Glossina* may promote the capacity of tsetse flies to transmit trypanosomes, the agent of sleeping sickness (Welburn and Maudlin 1999). If confirmed, strategies to depress *S. glossinidius* prevalence would reduce the vector competence of the tsetse fly.

In summary, there is now a sufficient research base to exploit symbioses in insects for novel pest management strategies. The priorities are to identify targets suitable for control, from analysis of the specific molecular interactions mediating transmission, nutritional interactions, the insect immune system, etc.; and to design strategies to alter symbiont abundance, varying from elimination of required symbionts to manipulation of symbionts that influence pest traits. These priorities relate directly to the increasing research interest in the processes underlying the persistence of symbioses (chapter 4, section 4.5), and the mechanisms by which hosts reward and sanction their partners (chapter 3, section 3.4.2).

5.5.3 *Symbiosis in Medicine and Human Health*

The resident microbiota on the multiple surfaces of the human body, including the skin, gut, and the respiratory, urinary, and reproductive tracts play a vital role in our well-being. They protect us from pathogens, promote immune function, contribute to our nutrition, and play an essential role in the development, particularly, of intestinal function. Imbalances in the composition and activity of the microbiota have also been implicated in some chronic diseases, notably in relation to the gut microbiota.

The evidence for microbial involvement in disease is particularly strong for irritable bowel syndrome, which comprises chronic inflammatory conditions affecting the entire gastrointestinal tract (Crohn's disease) or just the colon and rectum (ulcerative colitis). Mouse strains that are susceptible to colitis display symptoms in standard rearing conditions, but they are consistently asymptomatic when reared under sterile conditions isolated from all microorganisms (Elson et al. 2005); and many patients with the disease have a profile of microorganisms in the small intestine and colon that is significantly different from that of healthy people (Frank et al. 2007). Genetic approaches, including genome-wide association studies, have identified several susceptibility genes (e.g., *NOD2*) with a role in the innate immune response (Xavier and Podolsky 2007). Taken together, these data indicate firmly that these diseases are underlain by lesions in the interactions between the microbiota and immune system. This relates directly to the nature of the interactions between the immune system and the resident microbiota of mammals (see chapter 4, section 4.5.3).

There are indications that the composition of the gut microbiota can influence the progress of certain diseases, notably the frequently linked syndromes of insulin resistance (leading to type II diabetes) and obesity (Nicholson et al. 2005). Parallel research on humans and mice has demonstrated that the composition of the gut microbiota both influences and is influenced by obesity (Ley et al. 2006b; Turnbaugh et al. 2006). Genetically obese mice use food more efficiently than lean, wild-type mice. This effect can be attributed, at least partly, to the gut microorganisms because germ-free mice inoculated with the microbiota from obese mice accumulated more body fat than mice inoculated with microorganisms from lean mice (figure 5-10a). An indication of the complexity of the interaction between obesity and the gut microbiota comes from studies of humans. The gut microbiota of obese and lean people differ, as indicated by the higher ratio of Firmicutes to Bacteroidetes bacteria in stools of obese people. When obese people dieted for a year, the microbial composition of their stools became progressively more like that of lean people (figure 5-10b), and the people with the greatest weight reduction displayed the greatest change in bacterial profile. How do microbial communities characteristic of obese and lean people develop? Are there critical stages in development when diet, immune status, or other physiological traits set the conditions which favor different stable microbial communities? There is evidence that the overall structure of the gut microbial community develops in the first year of life (Palmer et al. 2007), potentially linking propensity to obesity with processes underlying the establishment of the symbiosis.

The medical importance of the gut microbiota extends beyond specific diseases. Gut microorganisms can have a large effect on the efficacy of orally administered drugs. Many drugs are inactivated in the gut by conjugation with bile salts, but microorganisms can extend the biological activity of the drug by hydrolyzing these conjugates to regenerate the active drug, which is then absorbed into the bloodstream. This is illustrated by research on a pharmacologically active plant flavonoid baicalein, which inhibits the growth of carcinoma cells. When rats were fed on a sugar conjugate of baicalein, the concentration of baicalein in the bloodstream was significantly lower in germ-free rats relative to untreated rats with gut microorganisms, as predicted (Akao et al. 2000).

As a result of the impact of gut microorganisms on drug delivery, symbiosis is emerging as an unexpected complicating factor in the application of personalized medicine. The aim of personalized medicine is to use our developing understanding of human genetic variation to prescribe drugs and other treatments tailored to individual genotypes as identified, for example, by single-nucleotide polymorphisms (SNPs). However, the response to drugs and other therapeutic interventions is not shaped exclusively by the human genome but also by the person's resident microbiota. The composition and activity of the microbiota are influenced by many environmental factors (e.g., diet, hygiene, history of infection, age, and physiological condition) and, consequently, the genotype of a person may not accurately predict the most appropriate medical treatment. We should anticipate that some aspects of personalized medicine will need to be personalized to the microbial symbiosis,

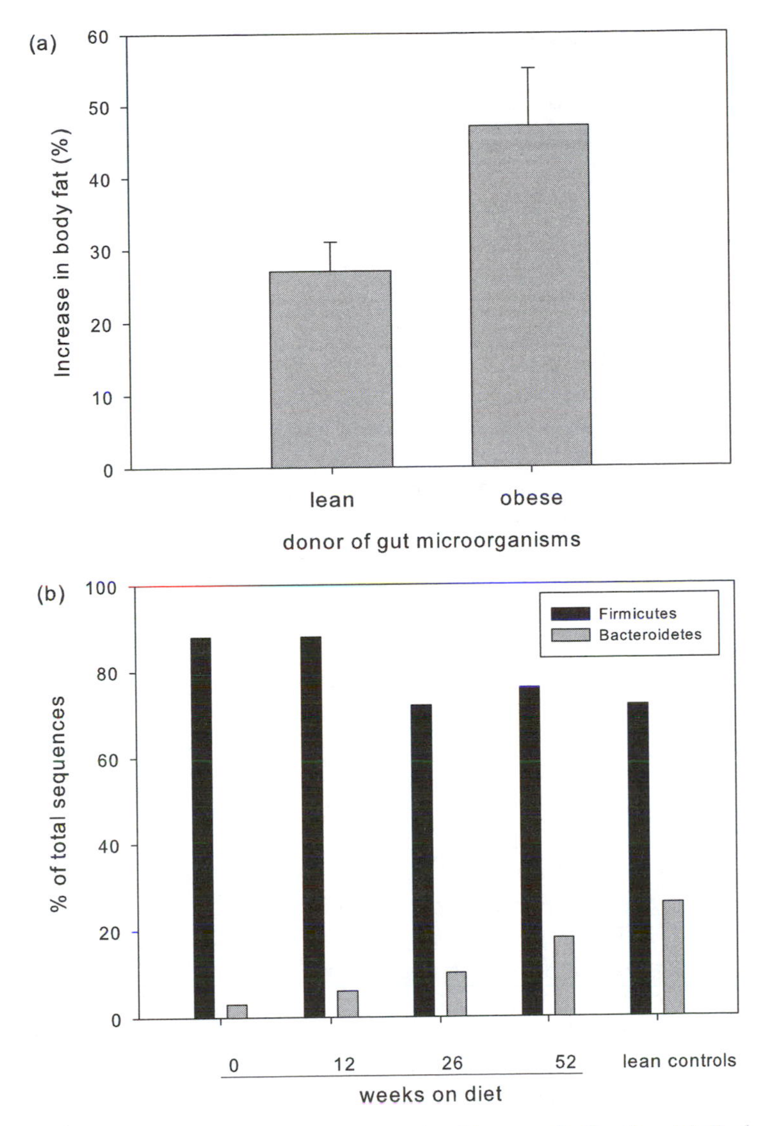

Figure 5-10 Interaction between gut microbiota and obesity. (a) Body fat content of germ-free mice colonized by microorganisms from the ceca of lean (wild-type) and genetically obese (*ob/ob*) mice. (b) Contribution of Firmicutes and Bacteroidetes bacteria to the stools of 12 obese people over one year of dieting. (These two groups of bacteria account for more than 90% of the total bacterial complement as determined from 16S rRNA gene sequencing.) [Adapted by permission from Macmillan Publishers, Ltd.: Nature, from figure 3c of Turnbaugh et al. (2006) and figure 1b of Ley et al. (2006b), copyright 2006]

which is apparently as unique as the human genotype (see chapter 4, section 4.3).

Linked to the evidence that human health and response to medical intervention are influenced by the resident microbiota, there is considerable interest in manipulating the composition and activity of microorganisms to alleviate disease and promote well-being. A further impetus for this approach comes from the recognition that we cannot continue to depend on antibiotics for the selective elimination of pathogens because of the increased incidence of antibiotic resistance among pathogens and antibiotic-mediated disruption of colonization resistance by the indigenous microbiota, leading to secondary infections.

Two alternative routes to manipulate the microbiota are being developed or are in use: administration of beneficial microorganisms, and identification of foods that promote a beneficial microbial community. Both are symbiotic problems, relating primarily to the evolutionary and physiological determinants of mutually beneficial interactions (chapters 2 and 3) and the factors that shape choice, competition in mixed infections, and persistence of symbioses (chapter 4).

In some contexts, beneficial microorganisms can be administered with the goal of out-competing pathogens and persisting with minimal impact on the host immune system. The nasopharynx and oral cavity are particularly accessible sites for such treatments.

Otitis media (earache) is caused by overgrowth of indigenous bacteria such as *Streptococcus pneumoniae*. Children prone to these ailments have low levels of streptococci with α-hemolytic activity which, in culture, inhibit the growth of *S. pneumoniae*. When children suffering from otitis media were treated with antibiotics and then administered α-hemolytic streptococci or a placebo by nasal spray, the incidence of recurrence of otitis media over three months was 22% for the experimental group and 42% for the control group (Roos et al. 2001). The bacterial spray provided no benefit in the absence of the antibiotic treatment, suggesting that, although the α-hemolytic streptococci are protective, they are not sufficiently competitive to displace *S. pneumoniae*.

Dental caries are caused by acid which is produced by bacterial fermentation, especially of dietary sugars, and then trapped at the tooth surface by extracellular polysaccharides also produced by the bacteria. One bacterium, *Streptococcus mutans*, releases particularly large amounts of acid, especially lactic acid, and polysaccharide. In principle, dental caries can be abolished by the replacement of *S. mutans* and other acidogenic bacteria by bacterial taxa that do not produce acids. The problem is that the nonacidogenic taxa are generally not competitive against the acidogenic forms (Tagg and Dierksen 2003), and the bacterial community in the oral cavity is very stable once established in very young children over the few months after onset of tooth eruption (Caulfield et al. 1993). One strategy is to obtain *S. mutans* variants which do not cause dental caries but which are highly competitive because they express antibiotics or bacteriocins active against other *S. mutans* strains. (The role of bacteriocins in competition among closely related bacteria is considered in chapter 4, section 4.4.1.) One strain genetically engineered for these traits has been demonstrated to protect rats against dental caries (Hillman 2002), but various issues relating to safety and persistence of the modified bacteria need to be addressed before application to humans can be considered (Tagg and Dierksen 2003).

Long-term persistence is not always the aim when live microorganisms are administered. Use of probiotics, i.e., the consumption of foods containing live microorganisms, is widely perceived to be beneficial, even though the ingested microorganisms rarely persist. The widely used bacteria in yoghurts, *Lactobacillus acidophilus* and *Brevibacterium breve*, are not known to be members of the indigenous gut microbiota and they are generally lost within a few days. Their presumptive beneficial effect is mediated primarily by the upregulation of the immune system, including activation of macrophages and increased production of IgA. The basis for the therapeutic value of one bacterium used as a probiotic has been established. The bacterium is the Nissle 1917 strain of *E. coli*, originally isolated from a First World War soldier who remained healthy through an outbreak of diarrhea. This strain has been used for many years in Central Europe, especially as protection from diarrhea caused by *Shigella* and *Salmonella*. Zyrek et al. (2007) have demonstrated that the Nissle 1917 strain promotes the formation and repair of tight junctions between intestinal epithelial cells, so creating a barrier to invasion by pathogenic bacteria.

A further strategy to obtain health benefits from beneficial microorganisms is the use of prebiotics. Prebiotics are foods intended to modify the overall community structure of the human microbiota. In particular, prebiotic oligofructans and inulins (long-chain fructans) are proposed to increase *Bifidobacterium* species and, thereby, to promote short-chain fatty acid production, improve mineral nutrition, and enhance immune function. The beneficial effects of these interventions appear to be greater for young children than adults (Seifert and Watzl 2007), consistent with other evidence that, once established, microbial communities in humans tend to be stable and resistant to change or invasion (Wilson 2005; Dethlefsen et al. 2007). Even so, there are some indications that dietary fructans can affect the activity of gut microorganisms in adults. One example relates to the isoflavone genistein, which is found in soyabeans and has anticancer properties but is prone to degradation by gut microorganisms. Dietary oligofructans can reduce this microbial activity, so protecting the potential therapeutic value of genistein (Steer et al. 2003).

In conclusion, the composition and activity of the resident microbiota have a profound impact on human physiology. The interactions are complex, involving interplay among microbial taxa and between the microbial community and the host. To manage the symbiosis to enhance human health, it is not sufficient to understand these interactions. It is also necessary to predict the impact of interventions (e.g., drugs, pre- or probiotics) on these interactions at the level of the individual human host of a unique genotype and microbial complement. The challenge is different from and greater than that posed by one-to-one interactions in highly specific symbioses, as in mycetocyte symbioses of insects or nitrogen-fixing symbioses in legumes. However, the task is not insuperable. Many of the resources needed to design predictive models of symbiosis function are available in the disciplines of community ecology and systems biology, and very large data sets on the microbiota and their metabolic and immunological impacts suitable for populating these models are being generated (Ley et al. 2006a)

5.6 Résumé

Symbiosis is an evolutionarily successful strategy by the criterion that it has promoted the diversification of various taxa. The acquisition of complex novel traits through symbiosis has opened up new ecological opportunities (resources, habitats, etc.) in which symbiotic organisms can diversify. For example, there is now persuasive evidence linking the diversification of ants to their transition from generalist scavengers to the exploitation, through various symbioses, of plants as a food source; and there is strong paleontological support for the role of photosynthetic symbiosis in the diversification of scleractinian corals (section 5.2.1). It has long been accepted that coevolutionary changes of the partners in symbioses can also lead to diversification through reproductive isolation and speciation. Nevertheless, the significance and processes underlying coevolutionary diversification are currently under review, with the increasing evidence that the reciprocal interactions in many associations are not played out in highly specific relationships between two partners but are diffuse, involving groups of similar species with similar benefits.

The ecological success of some symbioses can be attributed to antagonistic interactions between the symbiosis and other organisms. For example, the competitiveness of a symbiotic organism may be enhanced as a result of traits (e.g., improved nutrition) conferred by the association. The scale of these effects has been shown to be substantial for some plant-fungal relationships and, in general terms, symbiosis is likely to be ecologically significant where the association is not obligate and competing species vary in the benefit derived from symbioses. Importantly, the ecological consequence of symbiosis can vary: symbiosis can promote or depress biodiversity, and can promote biological invasions by exotic taxa or be vulnerable to such invasions.

In a world dominated by human activities, the success of symbiosis can also be gauged by their value to us. By this criterion, the potential success of symbiosis is high. Symbioses are being harnessed as sources of novel bioactive compounds. Other potential opportunities for intervention include the prevalence in humans of gut symbionts that influence the efficacy of medical drugs, and the incidence of bacteria that determine the pest status of certain insects. As we gain in our understanding of the symbiotic habit, so it will become increasingly feasible to predict and manage the impact of anthropogenic factors on ecologically important symbioses and to manipulate symbioses to enhance food production, pest control, and human health.

CHAPTER 6

Perspectives

THE CORE THEME of this book is that symbiosis, i.e., persistent mutually beneficial associations, is a pervasive habit among living organisms. Time and again, symbioses involving organisms with a variety of ancestral lifestyles have evolved, conferring advantage in habitats with limiting resources and in antagonistic interactions with predators, competitors, and pathogens (chapters 1 and 2). Despite these benefits, the organisms in symbioses regularly come into conflict over the allocation of resources. This conflict is managed by transmission mechanisms that promote the organisms' selective interest in the fitness of their partners and by rewards/sanctions that impose good behavior on the partners (chapter 3); and the incidence of conflict is also reduced by mechanisms for choosing cooperative partners and discriminating against partners that are ineffective or deleterious (chapter 4). These processes mediating the formation and persistence of the associations underlie the biological success of symbioses, as is illustrated by the role of symbiosis in triggering the diversification of various taxa and in structuring many ecological communities. There is also great potential to harness symbioses for enhanced food production, improved pest control, drug discovery, and the promotion of human health (chapter 5).

Despite the many advances in our understanding and appreciation of symbioses in recent years, we still have a very limited understanding of some aspects of the symbiotic habit, and I outline six key unresolved issues at the end of this final chapter. Importantly, our capacity to address these research problems is being enhanced greatly by recent technical developments, and I consider this issue first.

The context for the new opportunities afforded by technical developments is the traditional perspective that symbiosis research is difficult. Symbioses are inherently complex because they concern interactions between phylogenetically distant organisms. In addition, many symbioses cannot be maintained in the laboratory; and the participating organisms often have a long lifespan, are unsuitable for genetic methods, cannot be separated from their partners and grown in isolation, or lose their symbiotic traits when isolated. Furthermore, experiments are often limited by a paucity of biological material and small sample sizes. As a result, symbioses have often been regarded as intractable to analysis, especially in comparison to traditional model organisms, such as *E. coli*, *Drosophila*, and *Arabidopsis*, which are not generally considered as symbiotic organisms.

Increasingly, these traits of symbioses do not pose a barrier to successful research. Complex interactions are currently recognized as an important biological challenge, especially at the systems level for molecular to organismal interactions. Systems tools are available to symbiosis research. For example, Stolyar et al. (2007) have applied metabolic network analysis to explore the consortium between the sulfate-reducing bacterium *Desulfovibrio vulgaris* and the methanogen *Methanococcus maripalaudis*, in which the methanogen acts as an electron sink for its partner. Their metabolic model made specific predictions that could not have been obtained from traditional biochemical and genetic approaches concerning the identity and flux of metabolites between the partners and how the relative abundance of the two partners might be controlled. Subsequent experiments *in vivo* confirmed the predictions, including the expectation that hydrogen is more important than formate in electron transfer between the partners and that the biomass ratio of *D. vulgaris* to *M. maripalaudis* had to be at least 2:1 for the symbiosis to be stable.

The availability of molecular and genomic tools is making it less important to be able to cultivate symbiotic organisms. Organisms can be identified and genotyped from small DNA samples, even in complex multiorganismal mixtures. Massively parallel sequencing methods and ongoing developments for the analysis of metagenomes will also promote identification, evolutionary analysis, and functional studies. For example, function has been inferred from sequence in the analysis of the nutritional contributions of the two unculturable symbiotic bacteria, *Sulcia* and *Baumannia*, to their insect host, the glassy winged sharpshooter (McCutcheon and Moran 2007; see chapter 4, section 4.4.2 and figure 4-7). Proteomics and various gene expression strategies offer the opportunity to identify candidate genes and proteins underpinning function. Furthermore, routes to explore the symbiotic significance of genes identified by these methods are not restricted to organisms amenable to genetic approaches (e.g., for the production of knock-out mutants) because the expression of specific genes can be depressed in a wide range of eukaryotes by RNA interference. For example, RNAi of the caspase gene mediating apoptosis (cell death) of the sea anemone *Aiptasia pulchella* has been applied to establish that apoptosis plays a critical role in symbiosis breakdown and bleaching in response to elevated temperature (Dunn et al. 2007).

Symbiotic interactions which are lost on isolation can be explored using real-time microscopical methods applied to symbioses in situ. The power of these methods is illustrated by their contribution to the demonstration of the central role of calcium spiking in root hair cells of legume plants mediating infection by rhizobia symbionts and mycorrhizal fungi (chapter 4, section 4.2.4). Other analytical techniques are improving dramatically, enabling the detection and quantification of metabolites at low concentrations in small volumes and even in single cells.

As a result of these developments, it is becoming increasingly feasible to analyze biological systems, including symbioses, without the huge investment in the community tools and resources (databases, mutant banks, etc.) that is required to support research on the traditional model systems. Even so, for some purposes there are undoubted benefits for the research community to focus resources on single species representative of particular symbioses, especially where the genomes of the participating organisms have been sequenced and annotated in detail. These genomic resources are available, for example, to the symbiosis between the legume *Medicago truncatula* and *Sinorhizobium meliloti*, and between the pea aphid *Acyrthosiphon pisum* and bacterium *Buchnera*. Other emerging model systems with some genomic data include the symbioses between the sea anemone *Aiptasia* and *Symbiodinium* algae (Weis et al. 2008) and the bacterial symbiosis in the medicinal leech (Graf et al. 2006). There is also a great potential to tap the traditional model systems for symbiosis research. For example, the laboratory mouse is increasingly being used as a model for the impact of gut microbiota on human health (see chapter 5, section 5.5.3), and systems biology tools developed to construct and interrogate the metabolic network of the bacterium *E. coli* have been applied to explore the metabolism of symbiotic bacteria, e.g., *Buchnera*, whose genome is a subset of the *E. coli* genome (Thomas et al. 2009). In addition, *Arabidopsis* is proving suitable to study the mechanisms underlying the promotion of plant growth and seed production by endophytic fungi (e.g., Shahollari et al. 2007). Nevertheless, a traditional strength of symbiosis research has been the rich diversity of symbioses studied. It is important that the opportunities afforded by these emerging symbioitic models are matched by sustained research on other systems so that traits of general importance can be discriminated from those peculiar to one system.

These technical developments will prove very valuable in resolving six key questions for symbiosis research, as outlined below.

Can We Predict the Traits of Symbioses from the Ancestral Lifestyles of the Participating Organisms?

There is overwhelming evidence that the evolutionary origins of symbiotic organisms are diverse, including parasites, different symbioses, and free-living organisms; and some lineages may switch back and forth between symbiotic and free-living lifestyles (chapter 2). What are the implications for symbioses? For example, is conflict particularly evident in symbioses involving organisms with antagonistic ancestry, such that the incidence of cheaters is high and sanctions are severe? Is addiction (dependence without benefit) absent from associations involving lineages that switch between symbiotic and free-living states? Do the signaling modules mediating symbiosis formation and persistence tend to be derived from defense systems in symbioses with antagonistic ancestry and from developmental or regulatory signaling networks in associations involving organisms with symbiotic or free-living ancestries? Reliable data on the phylogenetic position and symbiotic traits of a wide diversity of symbiotic organisms are required to answer these questions. With the increasing availability of multigene and full genome data sets to obtain robust phylogenetic data for symbiotic organisms and their relatives, such analyses will increasingly yield reliable results.

When Is a Symbiont a Symbiont-Derived Organelle?

Symbiont-derived organelles have definitively evolved on just two occasions, giving rise to mitochondria and plastids. These organelles are of limited value in understanding the pattern of events at the evolutionary interface between symbionts and organelles because they evolved a billion or more years ago and have diversified subsequently (chapter 3, section 3.6). Recent discoveries provide new opportunities to explore the evolutionary transition from symbiont to organelle. Various vertically transmitted symbionts in protists and insects have very small genomes, and they are dependent on host products for many functions (chapter 3, section 3.6.5). Which traits, beyond their small genome, do these symbionts share with mitochondria and plastids? Have functions coded by host nuclear genes been allocated to the symbionts (i.e., subcontracting of function) possibly including the import of host proteins to the symbionts? Have functional genes been transferred from the symbiont to the host nucleus and the cognate proteins targeted either back to the symbiont or to other host compartments? Do the different symbionts with small genomes display similar combinations of traits? Such comparative analysis may reveal whether these traits evolve in a predictable order, providing insight into the processes underlying the transition from symbiont to organelle.

How Significant Are Cheaters as Determinants of the Traits of Symbioses?

The empirical evidence that organisms in symbioses cheat (i.e., derive advantage from being ineffective partners) demonstrates that conflict among partners is not resolved exclusively by the repeated, reciprocal exchange of services that are more valuable to the recipient than the donor (see chapter 3, sections 3.2.1 and 3.3). Various traits of symbioses that have been invoked to limit cheating and manage conflict include the recognition mechanisms which control access to the symbiosis, and the sanctions/rewards which promote the persistence of effective partners. However, these traits also protect the symbiosis from colonization by any ineffective organisms and provide a route to cull unhealthy partners, respectively. This raises the issue whether the recognition mechanisms function principally to exclude cheaters or more generally to discriminate against inappropriate and ineffective organisms. Similarly, are cheaters the principal target of sanctions/rewards in the established symbiosis, or do these enforcers of symbioses function primarily to remove those ineffective partners that, unlike cheaters, derive no advantage from their ineffectiveness? The answer to these questions is likely to vary across symbioses, depending on the type of symbiotic interactions, evolutionary origins of the partners, transmission mechanisms, etc. Of particular relevance is the incidence of cost-free traits (chapter 3, section 3.2.3) which, by definition, are not susceptible to cheating. Are cheaters more prevalent in symbioses with exclusively costly interactions than in those that have a mix of costly and cost-free interactions? Have some symbioses lost costly interactions and evolved to a cost-free basis as a route to evade cheaters? To gain a better grasp of the significance of cheaters in symbioses, we need to understand their incidence, the targets of recognition mechanisms and sanctions/rewards, and the distribution of cost-free traits.

Why Is Symbiosis Obligate for Some Organisms?

Some organisms are specialized for symbiosis with an apparently absolute inability to persist in isolation. This dependence is often regarded as a derived feature that has evolved in organisms with a long evolutionary history of symbiosis through the loss of traits that are redundant in symbiosis (or, in some instances, through genomic deterioration). However, dependence can also evolve rapidly by addiction without the context of mutual benefit (chapter 2, section 2.2.4). How widespread is addiction, and have some addictive relationships been misinterpreted as mutually beneficial? A linked issue is the mechanistic basis of obligacy. Among the few systems studied, the loss of redundant traits is mediated principally by gene loss, while addictive dependence is mediated by changes in the pattern of gene expression and signaling networks. Research on an array of obligate symbioses is needed to establish the validity of this apparent pattern and to understand how the molecular changes associated with addiction and loss of redundant traits affect the wider physiology, metabolism, and life-history traits of the participating organisms.

How Do Symbiotic Organisms Interact with the Immune/Defense System of Their Partners?

The defense systems (including the adaptive immune system of vertebrates) in animals and plants have traditionally been interpreted to function in the detection and elimination of foreign organisms, thereby conferring protection against pathogens and parasites. From this perspective, the harmless and beneficial organisms that coexist with these defense systems and sometimes even promote immune function are indicative of the limitations of the defense systems. This is an unsatisfactory interpretation. To accommodate symbioses, concepts of defense and immune systems are changing, with increasing interest in their role as surveillance systems that monitor and manage associated organisms. If correct, then the term defense system is perhaps a misnomer that has arisen from the research focus on pathogens; and defensive responses are just a small part of the role of systems that manage interactions with other organisms. The increasing recognition of the crucial importance of gut microorganisms to human health and the burgeoning research on interactions between animals and their gut microbiota provide the basis to resolve these important issues. Symbiosis has the potential to correct some basic principles of immunology.

Can Organisms Benefit from Symbiosis without Receiving a Service from Their Partner?

Some organisms benefit from an association, even though the effect cannot be attributed to a single defined product of the partner, such as a nutrient or protective chemical. Instead, some symbioses have a broad-based and beneficial impact on the hormonal, immune, metabolic, or other physiological system of participating organisms. It is as if the biology of symbiotic organisms is tuned to function well in the context of the symbiosis. For example, the nonpathogenic microbiota in an animal can prime the animal immune system, resulting in enhanced resistance to pathogens; and mycorrhizal or endophytic fungal infections can cause complex changes in the hormonal balance of plants, which are believed to promote plant resistance against pathogenic fungi and tolerance of drought and elevated temperatures. What are the traits of the partner triggering these changes, and are they orchestrated by a small number of master genes or by controls distributed across multiple regulatory networks? The global methods that are becoming increasingly available to analyze gene expression, signaling, and metabolic networks will be invaluable for resolving these issues.

I have every confidence that these questions will be resolved in the coming years. This will be facilitated not only by the technical developments outlined earlier in this chapter but also by the increasing confidence in the symbiosis research community. Symbiosis is now recognized as a discipline in its own right and its practitioners are no longer isolated into different disciplines (mycology, marine biology, microbiology, etc.). Symbiosis is not simply a type of interaction in the continuum of benefit and harm. It is a first-order process in the evolutionary diversification of living organisms, a crucial element to the physiological function of most eukaryotes, and a major determinant of the structure of ecological communities.

References

Akao, T., Kawabata, K., Yanagisawa, E., Ishihara, K., Mizuhara, Y., Wakui, Y., Sakashita, Y., and Kobashi, K. 2000. Balicalin, the predominant flavone glucuronide of scutellariae radix, is absorbed from the rat gastrointestinal tract as the aglycone and restored to its original form. *Journal of Pharmacy and Pharmacology* 52: 1563–1568.

Akiyama, K., Matsuzakik, I., and Hayashi, H. 2005. Plant sesquiterpenes induce hyphal branching in arbuscular mycorrhizal fungi. *Nature* 435: 824–827.

Akman, L., Yamashita, A., Watanabe, H., Oshima, K., Shiba, T., Hattori, M., and Aksoy S. 2002. Genome sequence of the endocellular obligate symbiont of tsetse flies, *Wigglesworthia glossinidia*. *Nature Genetics* 32: 402–407.

Allen, J. F. 2003. The function of genomes in bioenergetic organelles. *Philosophical Transactions of the Royal Society of London B* 358: 19–37.

Als, T. D., Villa, R., Kandul, N. P., Nash, D. R., Yen, S-H., Hsu, Y-F., Mignault, A. A., Roonsma, J. J., and Pierce, N. E. 2004. The evolution of alternative life histories in large blue butterflies. *Nature* 432: 386–390.

Althoff, D. M., Segraves, K. A., Leebens-Mack, J., and Pellmyr, O. 2006. Patterns of speciation in the yucca moths: Parallel species radiations within the *Tegeticula yuccasella* species complex. *Systematic Biology* 55: 398–410.

Anselme, C., Vallier, A., Balmand, S., Fauvarque, M-O., and Heddi, A. 2006. Host *PGRP* gene expression and bacterial release in endosymbiosis of the weevil *Sitophilus zeamais*. *Applied and Environmental Microbiology* 72: 6766–6772.

Ashburner, M., Hoy, M. A., and Peloquin, J. J. 1998. Prospects for the genetic transformation of arthropods. *Insect Molecular Biology* 7: 201–213.

Augé, R. M. 2001. Water relations, drought and vesicular-arbuscular mycorrhizal symbiosis. *Mycorrhiza* 11: 3–12.

Axelrod, R. 1984. *The Evolution of Co-operation*. Basic Books, New York.

Baco, A. R., and Smith, C. R. 2003. High species richness in deep-sea chemoautotrophic whale skeleton communities. *Marine Ecology Progress Series* 260: 109–114.

Baker, A. C., Starger, C. J., McClanahan, T. R., and Glynn, P. W. 2004. Corals' adaptive response to climate change *Nature* 430: 741–741.

Baldauf, S. L. 2003. The deep roots of eukaryotes. *Science* 300: 1703–1706.

Barth, F. G. 1991. *Insects and Flowers: The Biology of a Partnership*. Princeton University Press, Princeton, NJ.

Bascompte, J., and Jordano, P. 2007. Plant-animal mutualistic networks: the architecture of biodiversity. *Annual Reviews of Ecology, Evolution and Systematics* 38: 567–593.

Baylis, M., and Pierce, N. E. 1992. Lack of compensation by final instar larvae of the mycmecophilous lycaenid butterfly, *Jalmenus evagoras* for the loss of nutrients to ants. *Physiological Entomology* 17: 107–114.

Beier, M. S., Pumpini, C. B., Beier, J. C., and Davis, J. R. 1994. Effects of para-aminobenzoic acid, insulin, and gentamycin on *Plasmodium flaciparum* development in anopheline mosquitoes (Diptera: Culicidae). *Journal of Medical Entomology* 31: 561–565.

Bhavsar, A. P., Guttman, J. A., and Finlay, B. B. 2007. Manipulation of host-cell pathways by bacterial pathogens. *Nature* 449: 827–834.

Bidartondo, M. I., Burghardt, B., Gebauer, G., Bruns, T. D., and Read, D. J. 2004. Changing partners in the dark: isotopic and molecular evidence of ectomycorhizzal liaisons between forest orchids and trees. *Proceedings of the Royal Society of London B* 271: 1799–1806.

Bidartondo, M. I., Redecker, D., Hijri, I, Wiemken, A., Bruns, T. D., Dominguez, L., Sérsic, A., Leake, J. R., and Read, D. J. 2002. Epiparasitic plants specialized on arbuscular mycorrhizal fungi. *Nature* 419: 389–392.

Blanchard, J. L., and Lynch, M. 2000. Organellar genes—why do they end up in the nucleus? *Trends in Genetics* 16: 315–320.

Blaser, M. J., and Atherton, J. 2004. *Helicobacter pylori* persistence: biology and disease. *Journal of Clinical Investigation* 113: 321–333.

Blattner, F. R., Weising, K., Banfer, G., Maschwitz, U., and Fiala, B. 2001. Molecular analysis of phylogenetic relationships among myrmecophytic *Macaranga* species (Euphorbiaceae). *Molecular Phylogenetics and Evolution* 19: 331–344.

Boehm, T. 2006. Quality control in self/nonself discrimination. *Cell* 125: 845–858.

Bond, W. J. 1994. Do mutualisms matter? Assessing the impact of pollinator and disperser disruption on plant extinction. *Philosophical Transactions of the Royal Society of London B* 344: 83–90.

Brady, S. G., Schultz, T. R., Fisher, B. L., and Ward, P. S. 2006. Evaluating alternative hypotheses for the early evolution and diversification of ants. *Proceedings of the National Academy of Sciences USA* 103: 18172–18177.

Braendle, C., Miura, T., Bickel, R., Shingleton, A. W., Kambhampati, S., and Stern, D. L. 2003. Developmental origin and evolution of bacteriocytes in the aphid-*Buchnera* symbiosis. *PLoS Biology* 1: 70–76.

Brooks, M. A., and Richards, A. G. 1955. Intracellular symbiosis in cockroaches. II. Mitotic division of mycetocytes. *Science* 122: 242.

Brooks, T. M., Mittermeier, R. A., Mitermeier, C. G., da Fonseca, G.A.B., Rylands, A. B., Konstant, W. R., Flick, P., Pilgrim, J., Oldfield, S., Magin, G., and Hilton-Taylor, C. 2002. Habitat loss and extinction in the hotspots of biodiversity. *Conservation Biology* 16: 909–923.

Brouat, C., Garcia, N., Andary, C., and McKey, D. 2001. Plant lock and ant key: pairwise coevolution of an exclusion filter in an ant-plant mutualism. *Proceedings of the Royal Society of London B* 268: 2131–2141.

Brown, B. E., Dunne, R. P., Goodson, M. S., and Douglas, A. E. 2000. Bleaching patterns in reef corals. *Nature* 404: 142–143.

Brown, B. E., Downs, C. A., Dunne, R. P., and Gibb, S. W. 2002. Exploring the basis of thermotolerance in the reef coral *Goniastrea aspera*. *Marine Ecology Progress Series* 242: 119–129.

Brown, S. D. 1965. Chromosomal survey of the armoured and palm scale insects (Coccoidea: Diaspididae and Phoenicococcidae). *Hilgardia* 35: 189–294.

Brummel, T., Ching, A., Seroude, L., Simon, A. F., and Benzer, S. 2004. *Drosophila* lifespan enhancement by exogenous bacteria. *Proceedings of the National Academy of Sciences USA* 101: 12974–12979.

Brundrett, M. C. 2002. Coevolution of roots and mycorrhizas of land plants. *New Phytologist* 154: 275–304.

Bruns, T. D., and Shefferson, R. P. 2004. Evolutionary studies of ectomycorrhizal fungi: recent advances and future directions. *Canadian Journal of Botany* 82: 1122–1132.

Bryan-McKay, S. A., and Greeta, R. 2007. Protein subcellular relocalization: a new perspective on the origin of novel genes. *Trends in Ecology and Evolution* 22: 338–344.

Bryla, D. R., and Eissenstat, D. 2005. Respiratory costs of mycorrhizal associations. In *Plant Respiration. From Cell to Ecosystem,* ed. H. Lambers and M. Ribas-Carbo, 207–224. Springer, Dordrecht, Netherlands.

Bshary, R., and Grutter, A. S. 2005. Punishment and partner switching cause cooperative behaviour in a cleaning mutualism. *Biology Letters* 1: 396–399.

———. 2006. Image scoring and cooperation in a cleaner fish mutualism. *Nature* 441: 975–978.

Buchner, P. 1965. *Endosymbioses of Animals with Plant Microorganisms.* John Wiley, Chichester, UK.

Buckley, W. J., and Ebersole, J. P. 1994. Symbiotic organisms increase the vulnerability of a hermit crab to predation. *Journal of Experimental Marine Biology and Ecology* 182: 49–64.

Buddemeier, R. W., and Fautin, D. G. 1993. Coral bleaching as an adaptive mechanism: a testable hypothesis. *BioScience* 43: 320–326.

Bulgheresi, S., Schabussova, I., Chen, T., Mullin, N. P., Manzels, R. M., and Ott, J. A. 2006. A new C-type lectin similar to the human immunoreceptor DC-SIGN mediates symbiont acquisition by a marine nematode. *Applied and Environmental Microbiology* 72: 2950–2956.

Bull, J. J., Molineux, I. J., and Rice, W. R. 1991. Selection of benevolence in a host-parasite system. *Evolution* 45: 875–882.

Burdon, J. J., Gibson, A. H., Searle, S. D., Woods, M. J., and Brockwell, J. 1999. Variation in the effectiveness of symbiotic associations between native rhizobia and temperate Australian *Acacia*: within-species interactions. *Journal of Applied Ecology* 36: 398–408.

Burger, G., Gray, M. W., and Lang, B. F. 2003. Mitochondrial genomes: anything goes. *Trends in Genetics* 19: 709–716.

Callaway, R. M., Thelen, G. C., Rodriguez, A., and Holben, W. E. 2004. Soil biota and exotic plant invasion. *Nature* 427: 731–733.

Casiraghi, M., Bain, O., Guerrero, R., Martin, C., Pocacqua, C., Gardner, S. L., Franceschi, A., and Bandi, C. 2004. Mapping the presence of *Wolbachia pipientis* on the phylogeny of filarial nematodes: evidence for symbiont loss during evolution. *International Journal for Parasitology* 34: 191–203.

Caugant, D. A., Levin, B. R., and Selander, R. K. 1981. Genetic diversity and temporal variation in the *Escherichia coli* population of a human host. *Genetics* 98: 467–490.

Caulfield, P. W., Cutter, G. R., and Dasanayake, A. P. 1993. Initial acquisition of mutans streptococci by infants: evidence for a discrete window of infectivity. *Journal of Dental Research* 72: 37–45.

Cavalier-Smith, T. 1999. Principles of protein and lipid targeting in secondary symbiogenesis: euglenoid, dinoflagellate, and sporozoan plastid origins and the eukaryotic family tree. *Journal of Eukaryotic Microbiology* 46: 347–366.

Cesar, H., Burke, L., and Soede, L. P. 2003. The economics of worldwide coral reef degradation. A report compiled by *Cesar Environmental Consultancy* for WWF-Netherlands. Available at www.panda.org/coral.

Chao, L., Henley, K. A., Burch, C. L., Dahlberg, C., and Turner, P. E. 2000. Kin selection and parasite evolution: higher and lower virulence with hard and soft selection. *Quarterly Review of Biology* 75: 261.

Charlet, S., Hurst, G.D.D., and Merçot, H. 2003. Evolutionary consequences of *Wolbachia* infections. *Trends in Genetics* 19: 217–223.

Chen, M. C., Cheng, Y. M., Hong, M. C., and Fang, L. S. 2004. Molecular cloning of Rab5 (ApRab5) in *Aiptasia pulchella* and its retention in phagosomes harboring live zooxanthellae. *Biochemical and Biophysical Research Communications* 324: 1024–1033.

Cheney, K. L., and Coté, I. M. 2001. Are Caribbean cleaning symbioses mutualistic? Costs and benefits of visiting cleaning stations to longfin damselfish. *Animal Behaviour* 62: 927–933.

Christensen, B. B., Haagensen, J.A.J., Heydorn, A., and Molin, S. 2002. Metabolic commensalisms and competition in a two-species microbial consortium. *Applied and Environmental Microbiology* 68: 2495–2502.

Christian, C. E. 2001. Consequences of a biological invasion reveal the importance of mutualism for plant communities. *Nature* 413: 635–639.

Clarkson, T. W., Magos, L., and Myers, G. J. 2003. Human exposure to mercury: The three modern dilemmas. *Journal of Trace Elements in Experimental Medicine* 16: 321–343.

Clay, K., and Holah, J. 1999. Fungal endophyte symbiosis and plant diversity in successional fields. *Science* 285: 1742–1744.

Cleveland, L. R., Burke, A. W., and Karlson, P. 1960. Ecdysone-induced modifications in the sexual cycles of the protozoa of *Cryptocercus*. *Journal of Protozoology* 7: 229–239.

Coates, A. G., and Jackson, J.B.C. 1987. Clonal growth, algal symbiosis, and reef formation by corals. *Paleobiology* 13: 363–378.

Cole, S. T., and 44 coauthors. 2001. Massive gene decay in the leprosy bacillus. *Nature* 409: 1007–1011.

Conner, R. C. 1995. The benefits of mutualism: a conceptual framework. *Biological Reviews* 70: 427–457.

Cook, J. M., and Rasplus, J-Y. 2003. Mutualists with attitude: coevolving fig wasps and figs. *Trends in Ecology and Evolution* 18: 241–248.

Cooper, G., and Margulis, L. 1977. Delay in migration of symbiotic algae in *Hydra viridis* by inhibition of microtubule protein polymerisation. *Cytobios* 19: 7–19.

Cordeiro, N. J., and Howe, H. F. 2003. Forest fragmentation severs mutualism between seed dispersers and an endemic African tree. *Proceedings of the National Academy of Sciences USA* 100: 14052–14056.

Coté, I. M. 2000. Evolution and ecology of cleaning symbioses in the sea. *Oceanography and Marine Biology Annual Review* 38: 311–355.

Coyne, M. J., Reinap, B., Lee, M. M., and Comstock, E. 2005. Human symbionts use a host-like pathway for surface fucosylation. *Science* 307: 1778–1780.

Crutzen, P. J. 2002. Geology of mankind. *Nature* 415: 23.

Cullings, K. W. 1996. Single phylogenetic origin of ericoid mycorrhizae within the Ericaceae. *Canadian Journal of Botany* 74: 1896–1909.

Currie, C. R., Wong, B., Stuart, A. E., Schultz, T. R., Rehner, S. A., Mueller, U. G., Sung, G. H., Spatafora, J. W., and Straus, N. A. 2003. Ancient tripartite coevolution in the attine ant-microbe symbiosis. *Science* 299: 386–388.

Curtis, T. P., and Sloan, W. P. 2004. Prokaryotic diversity and its limits: microbial community structure in nature and implications for microbial ecology. *Current Opinion in Microbiology* 7: 221–226.

Dale, C., and Moran, N. A. 2006. Molecular interactions between bacterial symbionts and their hosts. *Cell* 126: 453–465.

Danforth, B. N. 1999. Emergence dynamics and bet hedging in a desert bee, *Perdita portalis*. *Proceedings of the Royal Society of London B* 266: 1985–1994.

Davidson, D. W., Cook, S. C., Snelling, R. R., and Chua, T. H. 2003. Explaining the abundance of ants in lowland tropical rainforest canopies. *Science* 300: 969–972.

Davidson, S. K., Allen, S. W., Lim, G. E., Anderson, C. M., and Haygood, M. G. 2001. Evidence for the biosynthesis of bryostatins by the bacterial symbiont "*Candidatus* Endobugula sertula" of the bryozoan *Endobugula neritina*. *Applied and Environmental Microbiology* 67: 4531–4537.

de Bruihn, F. J., Jing, Y., and Dazzo, F. B. 1995. Potential and pitfalls of trying to extend symbiotic interactions of nitrogen-fixing organisms to presently non-nodulated plants, such as rice. *Plant and Soil* 174: 225–240.

de Grey, A.D.N.J. 2005. Forces maintaining organellar genomes: is any as strong as genetic code disparity or hydrophobicity? *BioEssays* 27: 436–446.

de Souza, M. L., Newcombe, D., Alvey, S., Crowley, D. E., Hay, A., Sadowsky, M. J., and Wackett, L. P. 1998. Molecular basis of a bacterial consortium: interspecies catabolism of atrazine. *Applied and Environmental Microbiology* 64: 178–184.

Dethlefsen, L., McFall-Ngai, M., and Relman, D. A. 2007. An ecological and evolutionary perspective on human-microbe mutualism and disease. *Nature* 449: 811–818.

Dey, M., and Datta, S. K. 2002. Promiscuity of hosting nitrogen fixation in rice: An overview from the legume perspective. *Critical Reviews in Biotechnology* 22: 281–314.

Diez, J. 2005. Invasion biology of Australian ectomycorrhizal fungi introduced with eucalypt plantations into the Iberian Peninsula. *Biological Invasions* 7: 3–15.

Dodd, M. E., Silvertown, J., and Chase, M. 1999. Phylogenetic analysis of trait evolution and species diversity variation among angiosperm families. *Evolution* 43: 1308–1311.

Donia, M. S., Hathaway, B. J., Sudek, S., Haygood, M. G., Rosovitz, M. J., Ravel, K., and Schmidt, E. W. 2006. Natural combinatorial peptide libraries in cyanobacterial symbionts of marine ascidians. *Nature Chemical Biology* 2: 729–735.

Doolittle, W. F. 1998. You are what you eat: a gene transfer ratchet could account for bacterial genes in eukaryotic nuclear genomes *Trends in Genetics* 14: 307–311.

Douglas, A. E. 1983. Uric acid in *Platymonas convolutae* and symbiotic *Convoluta roscoffensis*. *Journal of the Marine Biological Association UK* 63: 435–447.

———. 1987. Specificity in the *Convoluta roscoffensis-Tetraselmis* symbiosis. In *Cell to Cell Signals in Plant, Animal and Microbial Symbiosis*, ed. S. Scannerini, D. C. Smith, P. Bonfante-Fasolo, and V. Gianinazzi-Pearson, 131–142. Elsevier, Amsterdam.

———. 1989. Mycetocyte symbiosis in insects. *Biological Reviews* 69: 409–434.

———. 1998. Nutritional interactions in insect-microbial symbioses. *Annual Reviews of Entomology* 43: 17–37.

———. 2003. Coral bleaching—how and why? *Marine Pollution Bulletin* 46: 385–392.

———. 2006. Phloem sap feeding by animals: problems and solutions. *Journal of Experimental Botany* 57: 747–754.

———. 2007. Symbiotic microorganisms: untapped resources for insect pest control. *Trends in Biotechnology* 25: 338–342.

———. 2008. Conflict, cheats and persistence of symbioses. *New Phytologist* 177: 849–858.

Douglas, A. E., and Smith, D. C. 1984. The green hydra symbiosis. VIII. Mechanisms in symbiont regulation. *Proceedings of the Royal Society of London B* 221: 291–319.

Douglas, S. E., Zauner, S., Fraunholz, M., Beaton, M., Penny, S., Deng, L-T., Wu, X., Reith, M., Cavalier-Smith, T., and Maier, U-G. 2001. The highly reduced genome of an enslaved algal nucleus. *Nature* 410: 1091–1096.

Dunn, S. R., Schnitzler, C. E., and Weis, V. M. 2007. Apoptosis and autophagy as mechanisms of dinoflagellate symbiont release during cnidarian bleaching: every which way you lose. *Proceedings of the Royal Society of London B* 274: 3079–3085.

Durvasula, R. V., Gumbs, A., Panackal, A., Kruglov, O., Aksoy, S., Merrifield, R. B., Richards, F. F., and Beard, C. B. 1997. Prevention of insect-borne disease: an approach using transgenic symbiotic bacteria. *Proceedings of the National Academy of Sciences USA* 94: 3274–3278.

Dyall, S. D., Koehler, C. M., Delgadillo-Correa, M. G., Bradley, P. J., Plümper, E., Leuenberger, D., Turck, C. W., and Johnson, P. J. 2000. Presence of a member of the mitochondrial carrier family in hydrogenosomes: conservation of membrane targeting pathways between hydrogenosomes and mitochondria. *Molecular and Cell Biology* 20: 2488–2497.

Eberle, M. W., and McLean, D. L. 1983. Observations of symbiote migration in human body lice with scanning and transmission electron microscopy. *Canadian Journal of Microbiology* 29: 755–762.

Eckburg, P. B., Bik, E. M., Bernstein, C. N., Purdom, E., Dethlefsen, L., Sargent, M., Gill, S. R., Nelson, K. E., and Relman, D. A. 2005. Diversity of the human intestinal microbial flora. *Science* 308: 1635–1638.

Elson, C. O., Cong, Y., McCracken, V. J., Dimmitt, R. A., Lorenz, R. G., and Weaver, C. T. 2005. Experimental models of inflammatory bowel disease reveal innate, adaptive, and regulatory mechanisms of host dialogue with the microbiota. *Immunology Reviews* 206: 260–276.

Eltz, T., Zimmermann, Y., Haftmann, J., Twele, R., Francke, W., Quezada-Euan, J.J.G., and Lunau, K. 2007. Enfleurage, lipid recycling and the origin of perfume collection in orchid bees. *Proceedings of the Royal Society of London B* 274: 2843–2848.

Embley, T. M., and Martin, W. 2006. Eukaryotic evolution, changes and challenges. *Nature* 440: 623–630.

Embley, T. M., van der Giezen, M., Horner, D. S., Dyal, P. L., and Foster, P. 2003. Mitochondria and hydrogenosomes are two forms of the same fundamental organelle. *Philosophical Transactions of the Royal Society of London B* 358: 191–202.

Ewald, P. W. 1994. *Evolution of Infectious Diseases*. Oxford University Press, Oxford.

Faeth, S. H., and Sullivan, T. J. 2003. Mutualistic asexual endophytes in a native grass are usually parasitic. *American Naturalist* 161: 310–325.

Farrell, B. D., Sequeira, B. C., and O'Meara, B. B. 2001. The evolution of agriculture in beetles (Curculionidae: Scolytinae and Platipodinae). *Evolution* 55: 2011–2027.

Fast, N. M., Kissinger, J. C., Roos, D. S., and Keeling, P. J. 2001. Nuclear-encoded, plastid targeted genes suggest a single common origin for apicomplexan and dinoflagellate plastids. *Molecular Biology and Evolution* 18: 418–426.

Fenchel, T., and Finlay, B. J. 1995. *Ecology and Evolution in Anoxic Worlds*. Oxford University Press, Oxford.

Fenn, K., and Blaxter, M. 2004. Are filarial nematode *Wolbachia* obligate mutualist symbionts? *Trends in Ecology and Evolution* 19: 163–166.

Fenner, F. 1983. Biological control, as exemplified by smallpox eradication and myxomatosis. *Proceedings of the Royal Society of London B* 218: 259–285.

Fitt, W. K., McFarland, F. K., Warner, M. E., and Chilcoat, G. C. 2000. Seasonal patterns of tissue biomass and densities of symbiotic dinoflagellates in reef corals and relation to coral bleaching. *Limnology and Oceanography* 45: 677–685.

Fitter, A. H. 2006. What is the link between carbon and phosphorus fluxes in arbuscular mycorrhizas? A null hypothesis for symbiotic function. *New Phytologist* 172: 3–6.

Flint, H. J., Duncan, S. H., Scott, K. P., and Louis, P. 2007. Interactions and competition within the microbial community of the human colon: links between diet and health. *Environmental Microbiology* 9: 1101–1111.

Frank, D. N., St. Amand, A. L., Feldman, R. A., Boedeker, E. C., Harpaz, N., and Pace, N. R. 2007. Molecular phylogenetic characterization of microbial community imbalances in human inflammatory bowel diseases. *Proceedings of the National Academy of Sciences USA* 104: 13780–13785.

Frank, S. A. 1996. Models of parasite virulence. *Quarterly Review of Biology* 71: 37–78.

Fulka, J., Fulka, H., St. John, J., Galli, C., Lazzasi, G., Lagutina, I., Fulka, J., and Loi, P. 2008. Cybrid human embryos—warranting opportunities to augment embryonic stem cell research. *Trends in Biotechnology* 26: 469–474.

Gabaldon, T., and Huynen, M. A. 2003. Reconstruction of proto-mitochondrial metabolism. *Science* 301: 609.

Gage, D. J. 2004. Infection and invasion of roots by symbiotic, nitrogen-fixing rhizobia during nodulation of temperate legumes. *Microbiology and Molecular Biology Reviews* 68: 280–300.

Gange, A. C., Brown, V. K., and Farmer, L. M. 1990. A test of mycorrhizal benefit in an early successional plant community. *New Phytologist* 115: 85–91.

Gaume, L., and McKey, D. 1999. An ant-plant mutualism and its host-specific parasite: activity rhythms, young leaf patrolling, and effects on herbivores of two specialist plant-ants inhabiting the same myrmecophyte. *Oikos* 84: 130–144.

Gherbi, H., Markmann, K., Svistoonoff, S., Estevan, J., Autran, D., Giczey, G., Auguy, F., Péret, B., Laplaze, L., Franche, C., Parniske, M., and Bogusz, D. 2008. SymRK defines a common genetic basis for plant root endosymbioses with AM fungi, rhizobia and Frankia bacteria. *Proceedings of the National Academy of Sciences USA* 105: 4928–4932.

Gilbert, L. E. 1980. In *Conservation Biology: an Evolutionary-Ecological Perspective*, ed. M. E., Soule and B. A. Wilcox, 11–33. Sinauer. Sunderland, MA.

Giraud, E., and 33 coauthors. 2007. Legumes symbioses: absence of *nod* genes in photosynthetic Bradyrhizobia. *Science* 316: 1307–1312.

Goreau, T. F. 1959. The physiology of skeleton formation in corals. I. A method for measuring the rate of calcium deposition by corals under different conditions. *Biological Bulletin* 116: 59–75.

Graf, J., Kikuchi, Y., and Rio, R.V.M. 2006. Leeches and their microbiota: naturally simple symbiosis models. *Trends in Microbiology* 14: 365–371.

Grant, V., and Grant, K. 1965. *Flower Pollination in the Phlox Family*. Columbia University Press, New York.

Gray, M. W., Burger, G., and Lang, B. F. 1999. Mitochondrial evolution. *Science* 281: 1476–1481.

Grime, J. P., Mackey, J.M.L., Hillier, S. H., and Read, D. J. 1987. Floristic diversity in a model system using experimental microcosms. *Nature* 328: 420–422.

Grutter, A. S., and Bshary, R. 2003. Cleaner wrasse prefer client mucus: support for partner control mechanisms in cleaning interactions. *Proceedings of the Royal Society of London B* 270: S244.

Gurdon, J. B., and Byrne, J. A. 2003. The first half-century of nuclear transplantation. *Proceedings of the National Academy of Sciences USA* 100: 8048–8052.

Hardin, G. 1968. The tragedy of the commons. *Science* 162: 1243–1248.

Hartmann, T. 1999. Chemical ecology of pyrrolizidine alkaloids. *Planta* 207: 483–495.

Hartnett, D. C., and Wilson, G.W.T. 1999. Mycorrhizae influence plant community structure and diversity in tall grass prairie. *Ecology* 80: 1187–1195.

Hauert, C., Holmes, M., and Doebeli, M. 2006. Evolutionary games and population dynamics: maintenance of cooperation in public goods games. *Proceedings of the Royal Society of London B* 273: 2565–2570.

Hauert, C., Traulsen, A., Brandt, H., Nowak, M. A., and Sigmund, K. 2007. Via freedom to coercion: The emergence of costly punishment. *Science* 316: 1905–1907.

Hause, B., Mrosk, C., Isayenkov, S., and Strack, D. 2007. Jasmonates in arbuscular mycorrhizal interactions. *Phytochemistry* 68: 101–110.

Hay, M. E., Parker, J. D., Burkepile, D. E., Caudill, C. C., Wilson, A. E., Hallinan, Z. P., and Chequer, A. D. 2004. Mutualisms and aquatic community structure: the enemy of my enemy is my friend. *Annual Review of Ecology, Evolution and Systematics* 35: 175–197.

Heil, M., and McKey, D. 2003. Protective ant-plant interactions as model systems in ecological and evolutionary research. *Annual Reviews of Ecology and Systematics* 34: 425–553.

Heil, M., Rattke, J., and Boland, W. 2005. Post-secretory hydrolysis of nectar sycrose and specialisation in ant/plant mutualism. *Science* 308: 560–563.

Henderson, B., Wilson, M., McNab, R., and Lax, A. J. 1999. *Cellular Microbiology: Bacteria-Host Interactions in Health and Disease*. Wiley, Chichester, UK.

Herre, E. A., Knowlton, N., Mueller, U. G., and Rehner, S. A. 1999. The evolution of mutualisms: exploring the paths between conflict and cooperation. *Trends in Ecology and Evolution* 14: 49–53.

Herrera, C.M. 2002. Seed dispersal by vertebrates. In *Plant-Animal Interactions*, ed. C. M. Herrera and O. Pellmyr, 185–208. Blackwell Science, Oxford.

Hillman, J. D. 2002. Genetically modified *Streptococcus mutans* for the prevention of dental caries. *Antonie Van Leeuwenhoek* 82: 361–366.

Hoegh-Guldberg, O. 1999. Climate change, coral bleaching and the future of the world's coral reefs. *Marine and Freshwater Research* 50: 839–866.

Hoegh-Guldberg, O., and 16 coauthors. 2007. Coral reefs under rapid climate change and ocean acidification. *Nature* 318: 1737–1742.

Hölldobler, B., and Wilson, E. O. 1990. *The Ants*. Harvard University Press, Cambridge, MA.

Honegger, R. 1984. Cytological aspects of the mycobiont-phycobiont relationship in lichens. *Lichenologist* 16: 111–127.

Hongoh, Y., and Ishikawa, H. 2000. Evolutionary studies on uricases of fungal endosymbionts of aphids and planthoppers. *Journal of Molecular Evolution* 51: 265–277.

Horn, M., and 12 coauthors. 2004. Illuminating the evolutionary history of chlamydiae. *Science* 304: 728–730.

Horner, D. S., Hirt, R. P., Kilvington, S., Lloyd, D., and Embley, T. M. 1996. Molecular data suggest an early acquisition of the mitochondrion endosymbiont. *Proceedings of the Royal Society of London B* 263: 1053–1059.

Hosokawa, T., Kikuchi, Y., Shimada, M., and Fukatsu, T. 2007. Obligate symbiont involved in pest status of host insect. *Proceedings of the Royal Society of London B* 274: 1979–1984.

Hotopp, J.C.D., and 20 coauthors. 2007. Widespread lateral gene transfer from intracellular bacteria to multicellular eukaryotes. *Science* 317: 1753–1756.

Howe, H. F., and Westley, L. C. 1988. *Ecological Relationships of Plants and Animals*. Oxford University Press, Oxford.

Hoy, M. A. 2000. Transgenic arthropods for pest management programs: Risks and realities. *Experimental and Applied Acarology* 24: 463–495.

Hu, L., and Kopecko, D. 1999. *Campylobacter jejuni* 81–176 associates with microtubules and dynein during invasion of human intestinal cells. *Infection and Immunity* 67: 4171–4182.

Hunter, M. S., Perlman, S. J., and Kelly, S. E. 2003. A bacterial symbiont in the Bacteroidetes induces cytoplasmic incompatibility in the parasitoid wasp *Encarsia pergandiella*. *Proceedings of the Royal Society of London B* 270: 2185–2190.

Hurst, G. D., and Jiggins, F. M. 2000. Male-killing bacteria in insects; mechanisms, incidence and implications. *Emerging Infectious Diseases* 6: 329–336.

Hurst, G.D.D., and Schilthuizen, M. 1998. Selfish genetic elements and speciation. *Heredity* 80: 2–8.

Huxley, C. 1980. Symbiosis between ants and epiphytes. *Biological Reviews* 55: 321–340.

Isack, H. A., and Reyer, H. U. 1989. Honey guides and honey gatherers—interspecies communication in a symbiotic relationship. *Science* 243: 1343–1346.

Izzo, T. J., and Vasconcelos, H. L. 2002. Cheating the cheater: domatia loss minimises the effects of ant castration in an Amazonian ant-plant. *Oecologia* 133: 200–205.

Jaenike, J., Dyner, K. A., Cornish, C., and Minhas, M. S. 2006. Asymmetric reinforcement and *Wolbachia* infection in *Drosophila*. *PLoS Biology* 4: e325.

Javot, H., Penmetsa, R. V., Terzaghi, N., Cook, D. R., and Harrison, M. J. 2007. A *Medicago truncatula* phosphate transporter indispensable for the arbuscular mycorrhizal symbiosis. *Proceedings of the National Academy of Sciences USA* 104: 1720–1725.

Jeon, K. W., and Ahn, T. I. 1978. Temperature sensitivity: a cell character determined by obligate endosymbionts in amoebas. *Science* 202: 635–637.

Jeon, T. J., and Jeon, K. W. 2003. Characterization of *sams* genes of *Amoeba proteus* and the endosymbiotic x-bacteria. *Journal of Eukaryotic Microbiology* 50: 61–69.

Johnson, M. D., Oldach, D., Delwiche, C. F., and Stoecker, D. K. 2007. Retention of transcriptionally active cryptophyte nuclei by the ciliate *Myrionecta rubra*. *Nature* 445: 426–428.

Johnstone, R. A., and Bshary, R. 2002. From parasitism to mutualism: partner control in asymmetric interactions. *Ecological Letters* 5: 634–739.

Jones, K. M., Kobayashi, H., Davies, B. W., Taga, M. E., and Walker, G. C. 2007. How rhizobial symbionts invade plants: the *Sinorhizobium-Medicago* model. *Nature Reviews Microbiology* 5: 619–633.

Joyce, S. A., Watson, R. J., and Clarke, D. J. 2006. The regulation of pathogenicity and mutualism in *Photorhabdus*. *Current Opinion in Microbiology* 9: 127–132.

Kaltenpoth, M., Göttler, W., Herzner, G., and Strohm, E. 2005. Symbiotic bacteria protect wasp larvae from fungal infestation. *Current Biology* 15: 475–479.

Karlberg, O., Canbäck, B., Kurland, C. G., and Andersson, S.G.E. 2000. The dual origin of the yeast mitochondrial proteome. *Yeast* 17: 170–187.

Kaufman, E., and Maschwitz, U. 2006. Ant-gardens of tropical Asian rainforests. *Naturwissenschaften* 93: 216–227.

Kellner, R.L.L., and Dettner, K. 1996. Differential efficacy of toxic pederin in deterring potential arthropod predators of *Paederus* (Coleoptera: Staphylinidae) offspring. *Oecologia* 107: 293–300.
Kiers, E. T., and van der Heijden, M.G.A. 2006. Mutualistic stability in the arbuscular mycorrhizal symbiosis: Exploring hypotheses of evolutionary cooperation. *Ecology* 87, 1627–1636.
Kiers, E. T., Rousseau, R. A., West, S. A., and Denison, R. F. 2003. Host sanctions and the legume-rhizobium mutualism. *Nature* 425: 78–81.
Kirkup, B. C., and Riley, M. A. 2004. Antibiotic-mediated antagonism leads to a bacterial game of rock-paper-scissors in vivo. *Nature* 428: 412–414.
Klasson, L., and Andersson, S.G.E. 2003. Evolution of minimal-gene-sets in host-dependent bacteria. *Trends in Microbiology* 12: 37–43.
Kneip, C., Lockhart, P., Voss, C., and Maier, U-G. 2007. Nitrogen fixation in eukaryotes—new models for symbiosis. *BMC Evolutionary Biology* 7: 55.
Knodler, L. A., Celli, J., and Finlay, B. B. 2001. Pathogenic trickery: Deception of host cell processes. *Nature Reviews Molecular Cell Biology* 2: 578–588.
Koga, R., Tsuchida, T., and Fukatsu, T. 2003. Changing partners in an obligate symbiosis: a facultative endosymbiont can compensate for loss of the essential endosymbiont *Buchnera* in an aphid. *Proceedings of the Royal Society of London B* 270: 2543–2550.
Kollock, P. 1998. Social dilemmas: the anatomy of cooperation. *Annual Review of Sociology* 24: 183–214.
Koropatnick, T. A., Engle, J. T., Apicella, M. A., Stabb, E. V., Goldman, W. E., and McFall-Ngai, M. J. 2004. Microbial factor-mediated development in a host-bacterial mutalism. *Science* 306: 1186–1188.
Koske, R. E., Gemma, J. N., and Englander, L. 1990. Vesicular-arbuscular mycorrhizae in Hawaiian Ericales. *American Journal of Botany* 77: 64–68.
Kosuta, S., Hazledine, S., Sun, J., Miwa, J., Morris, R. J., Downie, J. A., and Oldroyd, G.E.D. 2008. Differential and chaotic calcium signatures in the symbiosis signalling pathway of legumes. *Proceedings of the National Academy of Sciences USA* 105: 9823–9828.
Kurland, C. G., and Andersson, S.G.E. 2000. Origin and evolution of the mitochondrial proteome. *Microbiology and Molecular Biology Reviews* 64: 786–820.
Ladha, J. K., and Reddy, P. M. 2003. Nitrogen fixation in rice systems: state of knowledge and future prospects. *Plant and Soil* 252: 151–167.
Lafay, B., and Burdon, J. J. 1998. Molecular diversity of rhizobia occurring on native shrubby legumes in Southeastern Australia. *Applied and Environmental Microbiology* 64: 3989–3997.
LaJeunesse, T. C. 2002. Diversity and community structure of symbiotic dinoflagellates from Caribbean coral reefs. *Marine Biology* 141: 387–400.
Lawrence, J. G., Hendrix, R. W., and Casjens, S. 2001. Where are the pseudogenes in bacterial genomes? *Trends in Microbiology* 9: 535–540.
Lee, K. H., and Ruby, E. G. 1994. Competition between *Vibrio fischeri* strains during initiation and maintenance of a light organ symbiosis. *Journal of Bacteriology* 176: 1985–1991.

Lefèvre, C., Charles, H., Vallier, A., Delobel, B., Farrell, B., and Heddi, A. 2004. Endosymbiont phylogenesis in the Dryophthoridae weevils: evidence for bacterial replacement. *Molecular Biology and Evolution* 21: 965–973.

Letourneau, D. K. 1990. Code of ant-plant mutualism broken by parasite. *Science* 248: 215–217.

Lewis, D. H. 1985. Symbiosis and mutualism: crisp concepts and soggy semantics. In *The Biology of Mutualisms*, ed. D. H. Boucher, 29–39. Croom-Helm, London.

Ley, R. E., Peterson, D. A., and Gordon, J. I. 2006a. Ecological and evolutionary forces shaping microbial diversity in the human intestine. *Cell* 124: 837–848.

Ley, R. E., Turnbaugh, P. J., Klein, S., and Gordon, J. I. 2006b. Microbial ecology—Human gut microbes associated with obesity. *Nature* 444: 1022–1023.

Lo, N., Bandi, C., Watanabe, H., Nalepa, C., and Beninati, T. 2003. Evidence for cocladogenesis between diverse dictyopteran lineages and their intracellular endosymbionts. *Molecular Biology and Evolution* 20: 907–913.

Lo, N., Tokuda, G., Watanabe, H., Rose, H., Slaytor, M., Maekawa, K., Bandi, C., and Noda, H. 2000. Evidence from multiple gene sequences indicates that termites evolved from wood-feeding cockroaches. *Current Biology* 10: 801–804.

Lunn, J. E. 2002. Evolution of sucrose synthesis. *Plant Physiology* 128: 1490–1500.

Lutzoni, F., Pagel, M., and Reeb, V. 2001. Major fungal lineages are derived from lichen symbiotic ancestors. *Nature* 411: 937–940.

Machado, C. A., Robbins, N., Gilbert, M.T.P., and Herre, E. A. 2005. Critical review of host specificity and its coevolutionary implications in the fig/fig-wasp mutualism. *Proceedings of the National Academy of Sciences USA* 102: 6558–6565.

Manuel, R. 1988. *British Anthozoa*. Linnean Society of London, London.

Marie, C., Broughton, W. J., and Deakin, W. J. 2001. *Rhizobium* type III secretion systems: legume charmers or alarmers? *Current Opinion in Plant Biology* 4: 336–342.

Marin, B., Nowack, E.C.M., and Melkonian, M. 2005. A plastid in the making: primary endosymbiosis. *Protist* 156: 425–432.

Marjerus, M., Amos, W., and Hurst, G. 1996. *Evolution: The Four Billion Year War*. Longman, Harlow, UK.

Markmann, K., Giczey, G., and Parniske, M. 2008. Functional adaptation of a plant receptor kinase paved the way for the evolution of intracellular root symbioses with bacteria. *PLoS Biology* 6: 497–506.

Marler, M. M., Zabinski, C. A., and Callaway, R. M. 1999. Mycorrhizae indirectly enhance competitive effects of an intensive forb on a native bunchgrass. *Ecology* 80: 1180–1186.

Martin, F., Kohler, A., and Duplessis, S. 2007. Living in harmony in the wood underground: ectomycorrhizal genomics. *Current Opinion in Plant Biology* 10: 204–210.

Martin, W. 2005. Archaebacteria (Archaea) and the origin of the eukaryotic nucleus. *Current Opinion in Microbiology* 8: 630–637.

Martinez, E., Palacois, R., and Sanchez, F. 1987. Nitrogen-fixing nodules induced by *Agrobacterium tumefaciens* harboring *Rhizobium phaseoli* plasmids. *Journal of Bacteriology* 169: 2828–2834.

Marussich, W. A., and Machado, C. A. 2007. Host-specificity and coevolution among pollinating and nonpollinating New World fig wasps. *Molecular Ecology* 16: 1925–1946.

Maschwitz, U., and Hanel, H. 1985.The migrating herdsman *Dolichoderus (Diabolus) cuspidatus*—an ant with a novel mode of life. *Behavioural Ecology and Sociobiology* 17: 171–184.

Matzinger, P. 2002. The danger model: a renewed sense of self. *Science* 296: 301–305.

Maynard Smith, J., and Harper, D. 2003. *Animal Signals*. Oxford University Press, Oxford.

Maynard Smith, J. and Szmathary, E. 1995. *The Major Transitions in Evolution*. Oxford University Press, Oxford.

Mayr, E. 1960. The emergence of evolutionary novelties. In *Evolution after Darwin*, ed. S. Tax, 42–79. Chicago University Press, Chicago.

McAuley, P. J. 1986. The cell cycle of symbiotic *Chlorella*. III. Numbers of algae in green hydra digestive cells are regulated at digestive cell division. *Journal of Cell Science* 85: 63–71

McCutcheon, J. P., and Moran, N. A. 2007. Parallel genomic evolution and metabolic interdependence in an ancient symbiosis. *Proceedings of the National Academy of Sciences USA* 104: 19392–19397.

McFall-Ngai, M. 2007. Care for the community. *Nature* 445: 153.

McGovern, T. M., and Hellberg, M. E. 2003. Cryptic species, cryptic endosymbionts, and geographical variation in chemical defences in the bryozoan *Bugula neritina*. *Molecular Ecology* 12: 1207–1215.

McLean, C. B., Cunningham, J. H., and Lawrie, A. C. 1999. Molecular diversity within and between ericoid endophytes from the Ericaceae and Epacridaceae. *New Phytologist* 144: 351–358.

Medzhitov, R., and Janeway, C. A. 2002. Decoding the patterns of self and nonself by the innate immune system. *Science* 296: 298–300.

Mira, A., and Moran, N. A. 2002. Estimating population size and transmission bottlenecks in maternally transmitted endosymbiotic bacteria. *Microbial Ecology* 44: 137–143.

Modjo, H. S., and Hendrix, J. W. 1986. The mycorrhizal fungus *Glomus macrocarpum* as a cause of tobacco stunt disease. *Phytopathology* 76: 688–691.

Molina, R., Massicotte, H., and Trappe, J. M. 1992. Specificity phenomena between mycorrhizal symbioses; community-ecological consequences and practical implications. In *Mycorrhizal Functioning*, ed. M. F. Allen, 357–423. Chapman and Hall, New York.

Molmeret, M., Horn, M., Wagner, M., Santic, M., and Abu Kwaik, Y. 2005. Amobae as training grounds for intracellular bacterial pathogens. *Applied and Environmental Microbiology* 71: 20–28.

Montllor, C. B., Maxmen, A., and Purcell, A. H. 2002. Facultative bacterial endosymbionts benefit pea aphids *Acyrthosiphon pisum* under heat stress. *Ecological Entomology* 27: 189–195.

Moon, C. D., Craven, K. D., Leuchtmann, A., Clement, S. L., and Schardl, C. L. 2004. Prevalence of interspecific hybrids among asexual fungal endophytes of grasses. *Molecular Ecology* 13: 1455–1467.

Mooney, K. A. 2006. the disruption of an ant-aphid mutualism increases the effects of birds on pine herbivores. *Ecology* 87: 1805–1815.

Moran, N. A. 1996. Accelerated evolution and Muller's rachet in endosymbiotic bacteria. *Proceedings of the National Academy of Sciences USA* 93: 2873–2878.

Moran, N. A., and Wernegreen, J. J. 2000. Lifestyle evolution in symbiotic bacteria: insights from genomics. *Trends in Ecology and Evolution* 15: 321–326.

Moran, N. A., Degnan, P. H., Santos, S. R., Dunbar, H. E., and Ochman, H. 2005. The players in a mutualistic symbiosis: Insects, bacteria, viruses, and virulence genes. *Proceedings of the National Academy of Sciences USA* 102: 16919–16926.

Moran, N. A., Munson, M. A., Baumann, P., and Ishikawa, H. 1993. A molecular clock in endosymbiotic bacteria is calibrated using the insect hosts. *Proceedings of the Royal Society of London B* 253: 167–171.

Moran, N. A., Tran, P., and Gerardo, N. M. 2005. Symbiosis and insect diversification: an ancient symbiont of sap-feeding insects from the bacterial phylum *Bacteroidetes*. *Applied and Environmental Microbiology* 71: 8802–8810.

Moulin, L., Munive, A., Dreyfus, B., and Boivin-Masson, C. 2001. Nodulation of legumes by members of the β-subclass of proteobacteria. *Nature* 411: 948–950.

Moya, A., Tambutté, S., Tambutté, E., Zoccola, D., Caminiti, N., and Allemand, D. 2006. Study of calcification during a daily cycle of the coral *Stylophora pistillata*: implications for "light-enhanced calcification." *Journal of Experimental Biology* 209: 3413–3419.

Muller, H. J. 1964. The relation of recombination to mutational advance. *Mutation Research* 1: 2–9.

Mullins, D. E., and Cochran, D. G. 1975. Nitrogen metabolism in American cockroach. 2. Examination of negative nitrogen balance with respect to mobilization of uric acid stores. *Comparative Biochemistry and Physiology* 50: 501–510.

Nakabachi, A., Yamashita, A., Toh, H., Ishikawa, H., Dunbar, H. E., Moran, N. A., and Hattori, M. 2006. The 160-kilobase genome of the bacterial endosymbiont *Carsonella*. *Science* 314: 267.

Nash, D. R., Als, T. D., Maile, R., Jones, G. R., and Boomsma, J. J. 2008. A mosaic of chemical coevolution in a large blue butterfly. *Science* 319: 88–90.

Nault, L. R., and Rodriguez, J. G. 1985. *The Leafhoppers and Planthoppers*. Wiley InterScience, New York.

Newman, E. I., and Reddell, P. 1987. The distribution of mycorrhizas among families of vascular plants. *New Phytologist* 106: 745–751.

Newsham, K. K., Fitter, A. H., and Watkinson, A. R. 1995. Arbuscular mycorrhiza protect an annual grass from root pathogenic fungi in the field. *Journal of Ecology* 83: 991–1000.

Nicholson, J. K., Holmes, E., and Wilson, I. D. 2005. Gut microorganisms, mammalian metabolism and personalized health care. *Nature Reviews Microbiology* 3: 431–438.

Nikoh, N., Tanaka, K., Shibata, F., Kondo, N., Hizume, M., Shimada, M., and Fukatsu, T. 2008. Wolbachia genome integrated in an insect chromosome: evolution and fate of laterally transferred endosymbiont genes. *Genome Research* 18: 272–280.

Noda, H., and Koizumi, Y. 2003. Sterol biosynthesis by symbiotes: cytochrome P450 sterol C-22 desaturase genes from yeastlike symbiotes of rice plant-hoppers and anobiid beetles. *Insect Biochemistry and Molecular Biology* 33: 649–658.

O'Brien, H. E., Miadlikowska, J., and Lutzoni, F. 2005. Assessing host specialization in symbiotic cyanobacteria associated with four closely related species of the lichen fungus *Peltigera*. *European Journal of Phycology* 40: 363–378.

Offenberg, J. 2001. Balancing between mutualism and exploitation: the symbiotic interaction between *Lasius* ants and aphids. *Behavioural Ecology and Sociobiology* 49: 304–310.

Pallen, M. J., and Wren, B. W. 2007. Bacterial pathogenomics. *Nature* 449: 835–842.

Palmer, C., Bik, E. M., DiGiulio, D. B., Relman, D. A., and Brown, P. O. 2007. Development of the human infant intestinal microbiota. *PLoS Biology* 5: e177.

Palmer, T. M. 2004. Wars of attrition: colony size determines competitive outcomes in a guild of African acacia ants. *Animal Behaviour* 68: 993–1004.

Pannebakker, B. A., Loppin, B., Elemans, C.P.H., Humblot, L., and Vavre, F. 2007. Parasitic inhibition of cell death facilitates symbiosis. *Proceedings of the National Academy of Sciences USA* 104: 213–215.

Pellmyr, O. 2002. Pollination by animals. In *Plant-Animal Interactions*, ed. C. M. Herrera and O. Pellmyr, 157–184. Blackwell Scientific, Oxford.

Perez-Brocal, V., Gil, R., Ramos, S., Lamelas, A., Postigo, M., Michelena, J. M., Silva, F. J., Moya, A., and Latorre, A. 2006. A small microbial genome: The end of a long symbiotic relationship? *Science* 314: 312–313.

Perrine-Walker, F. M., Prayitno, J., Rolfe, B. G., Weinman, J. J., and Hocart, C. H. 2007. Infection process and the interaction of rice roots with rhizobia. *Journal of Experimental Biology* 58: 3343–3350.

Pham, L. N., Dionne, M. S., Shirasu-Hiza, M., and Schneider, D. S. 2007. A specific primed immune response in *Drosophila* is dependent on phagocytes. *PLoS Pathogens* 3: e26.

Piel, J. 2002. A polyketide synthase-peptide synthetase gene cluster from an uncultured bacterial symbiont of *Paederus* beetles. *Proceedings of the National Academy of Sciences USA* 99: 14002–14007.

Piel, J., Hui, D. Q., Wen, G. P., Butske, D., Platzer, M., Fusetani, N., and Matsunaga, S. 2004. Antitumor polyketide biosynthesis by an uncultivated bacterial symbiont of the marine sponge *Theonella swinhoei*. *Proceedings of the National Academy of Sciences USA* 101: 16222–16227.

Pierce, N. E., Braby, M. F., Heath, A., Lohman, D. J., Mathew, J., Rand, D. B., and Travassos, M. A. 2002. The ecology and evolution of ant association in the Lycaenidae (Lepidoptera). *Annual Reviews of Entomology* 47: 733–771.

Pirozynski, K. A., and Malloch, D. W. 1975. The origin of land plants: a matter of mycotropism. *BioSystems* 6: 153–164.

Pochon, X., Montoya-Burgos, J. I., Stadelmann, B., and Pawlowski, J . 2006. Molecular phylogeny, evolutionary rates, and divergence timing of the symbiotic dinoflagellate genus *Symbiodinium*. *Molecular Phylogenetics and Evolution* 38: 20–30.

Pumpuni, C. B., Beier, M. S., Nataro, J. P., Guers, L. D., and Davis, J. R. 1993. *Plasmodium flaciparum*: inhibition of sporogonic development in *Anopheles stepensi* by Gram-negative bacteria. *Experimental Parasitology* 77: 195–199.

Redman, R. S., Davigan, D. D., and Rodriguez, R. J. 2001. Fungal symbiosis from mutualism to parasitism: who controls the outcome, host or invader? *New Phytologist* 151: 705–716.

Redman, R. S., Sheehan, K. B., Stout, R. G., Rodriguez, R. J., and Henson, J. M. 2002. Thermotolerance generated by plant/fungal symbiosis. *Science* 298: 1581.

Reeve, J. N. 1999. Archaebacteria then . . . archaes now (are there really no archaeal pathogens?). *Journal of Bacteriology* 181: 3613–3617.

Richardson, D. M., Allsopp, N., D'Antonio, C. M., Milton, S. J., and Rejmanek, M. 2000. Plant invasions—the role of mutualisms. *Biological Reviews* 75: 65–93.

Rikkinen, J., Oksanen, I., and Lohtander, K. 2002. Lichen guilds share related cyanobacterial symbionts. *Science* 297: 357–357.

Riley, A. D., and Gordon, D. 1999. The ecological role of bacteriocins. *Trends in Microbiology* 7: 129–133.

Rillig, M. C. 2004. Arbuscular mycorrhizae, glomalin, and soil aggregation. *Canadian Journal of Soil Science* 84: 355–363.

Rocha, E.P.C., and Danchin, A. 2002. Base composition bias might result from competition for metabolic resources. *Trends in Genetics* 18: 291–294.

Rodriguez-Lanetty, M., Phillips, W. S., and Weis, V. M. 2006. Transcriptome analysis of a cnidarian-dinoflagellate mutualism reveals complex modulation of host gene expression. *BMC Genomics* 7: 23.

Roos, K., Grahn Håkansson, E., and Holm, S. 2001. Effect of recolonisation with "interfering" α streptococci on recurrences of acute and secretory otitis media in children: Randomised placebo controlled trial. *British Medical Journal* 322: 210–212.

Ross, D. M. 1971. Protection of hermit crabs (*Dardanus* spp.) from octopus by commensal sea anemones (*Calliactis* spp.). *Nature* 230: 401–402.

Roth, L. E., Jeon, K., and Stacey, G. 1988. Homology in endosymbiotic systems: that term "symbiosome." In *Molecular Genetics of Plant-Microbe Interactions*. ed. R. Palacios and D.P.S. Verma, 220–225. APS Press, St Paul, MN.

Rowland, I. R. 1999. Toxicological implications of the normal microbiota. In *Medical Importance of the Normal Microflora*, ed. G. W. Tannock, 295–311. Kluwer Academic Publishers, Dordrecht.

Ruby, E. G., and Asato, L. M. 1993. Growth and flagellation of *Vibrio fischeri* during initiation of the sepiolid squid light organ symbiosis. *Archives of Microbiology* 159: 160–167.

Ruddiman, W. F. 2005. *Plows, Plagues and Petroleum*. Princeton University Press, Princeton, NJ.

Rudgers, J. A., Holah, J., Orr, S. P., and Clay, K. 2007. Forest succession suppressed by an introduced plant-fungal symbiosis. *Ecology* 88: 18–25.

Rumpho, M. E., Worful, J. M., Lee, J., Kannan, K., Tyler, M. S., Bhattacharya, D., Moustafa, A., and Manhart, J. R. 2008. Horizontal gene transfer of the algal nuclear gene *pbsO* to the photosynthetic sea slug *Elysia chlorotica*. *Proceedings of the National Academy of Sciences USA* 105: 17867–17871.

Ryan, F. 2003. *Darwin's Blind Spot*. Thomson and Thomson, London.

Sachs, J. L., and Simms, E. L. 2006. Pathways to mutualism breakdown. *Trends in Ecology and Evolution* 21: 585–592.

Sachs, J. L., and Wilcox, T. P. 2006. A shift to parasitism in the jellyfish symbiont *Symbiodinium microadriaticum*. *Proceedings of the Royal Society of London B* 273: 425–429.

Sachs, J. L., Mueller, U. G., Wilcox, T. P., and Bull, J. J. 2004. The evolution of cooperation. *Quarterly Review of Biology* 79: 135–160.

Saldarriaga, J. F., Taylor, F.J.R., Keeling, P. J., and Cavalier-Smith, T. 2001. Dinoflagellate nuclear SSU rRNA phylogeny suggests multiple plastid losses and replacements. *Journal of Molecular Evolution* 53: 204–213.

Santos, S. R., Taylor, D. J., and Coffroth, M. A. 2001. Genetic comparisons of freshly isolated versus cultured symbiotic dinoflagellates: Implications for extrapolating to the intact symbiosis. *Journal of Phycology* 37: 900–912.

Sapp, J. 1994. *Evolution by Association*. Oxford University Press, Oxford.

Sauer, C., Stackebrandt, E., Gadau, J., Holldobler, B., and Gross, R. 2000. Systematic relationships and cospeciation of bacterial endosymbonts and their carpenter ant host species: proposal of the new taxon Candidatus *Blochmannia* gen. nov. *International Journal of Systematic and Evolutionary Microbiology* 50: 1877–1886.

Schardl, C. L. 1996. *Epichloë* species: fungal symbionts of grasses. *Annual Reviews of Phytopathology* 34: 109–130.

Schüssler, A., and Kluge, M. 2000. *Geosiphon pyriforme*, an endocytosymbiosis between fungus and cyanobacteria, and its meaning as a model system for arbuscular mycorrhizal research. In *The Mycota IX Fungal Associations*, ed. B. Hock, 151–161. Springer-Verlag, Berlin.

Schwartz, M. W., Hoeksema, J. D., Gehring, C. A., Johnson, N. C., Klironomos, J. N., Abbott, L. K., and Pringle, A. 2006. The promise and potential consequences of the global transport of mycorrhizal fungal inoculum. *Ecology Letters* 9: 501–515.

Seifert, S., and Watzl, B. 2007. Inulin and oligofructose: Review of experimental data on immune modulation. *Journal of Nutrition* 137: 2563S–2567S.

Seong, S-Y., and Matzinger, P. 2004. Hydrophobicity: an ancient damage-associated molecular pattern that initiates innate immune responses. *Nature Reviews Immunology* 4: 469–477.

Shahollari, B., Vadassery, J., Varma, A., and Oelmuller, R. 2007. A leucine-rich repeat protein is required for growth promotion and enhanced seed production mediated by the endophytic fungus *Piriformospora indica* in *Arabidopsis thaliana*. *Plant Journal* 50: 1–13.

Sheppard, C.R.C. 2003. Predicted recurrences of mass coral mortality in the Indian Ocean. *Nature* 425: 294–297.

Shigenobu, S., Watanabe, H., Hattori, M., Sasaki, Y., and Ishikawa, H. 2000. Genome sequence of the endocellular bacterial symbiont of aphids *Buchnera* sp. APS. *Nature* 407: 81–86.

Shingleton, A. W., Stern, D. L., and Foster, W. A. 2005. The origin of a mutualism: A morphological trait promoting the evolution of ant-aphid mutualisms. *Evolution* 59: 921–926.

Silver, A. C., Kikuchi, Y., Fadl, A. A., Sha, J., Chopra, A. K., and Graf, J. 2007. Interaction between innate immune cells and a bacterial type III secretion system in mutualistic and pathogenic associations. *Proceedings of the National Academy of Sciences USA* 104: 9481–9486.

Simon, L., Bousquet, J., Levesque, R., and Lalonde, M. 1993. Origin and diversification of endomycorrhizal fungi and coincidence with vascular land plants. *Nature* 363: 67–69.

Simser, J. A., Rahman, M. S., Dreher-Lesnick, S. M., and Azad, A. F. 2005. A novel and naturally occurring transposon, ISRpe in the *Rickettsia peacockii* genome disrupting the *rickA* gene involved in actin-based motility. *Molecular Microbiology* 58: 71–79.

Smith, D. C. 1979. From extracellular to intracellular: establishment of a symbiosis. *Proceedings of the Royal Society of London B* 204: 115–130.

Smith, S. E., and Read, D. J. 2007. *Mycorrhizal Symbiosis*. Academic Press, San Diego.

Stachowicz, J. J., and Hay, M. E. 1999. Mutualism and coral persistence: the role of herbivore resistance to algal chemical defence. *Ecology* 80: 2085–2101.

Stachowicz, J. J., and Whitlach, R. B. 2005. Multiple mutualists provide complementary benefits to their seaweed hosts. *Ecology* 86: 2418–2427.

Stadler, B., and Dixon, A.F.G. 2005. Ecology and evolution of aphid-ant interactions. *Annual Review of Ecology, Evolution and Systematics* 36: 345–372.

Stanley, G. D. 2003. The evolution of modern corals and their early history. *Earth Science Reviews* 60: 195–225.

Stanley, G. D., and Swart, P. K. 1995. Evolution of the coral-zooxanthellae symbiosis during the Triassic: a geochemical approach. *Paleobiology* 21: 179–199.

Stanton, M. L. 2003. Interacting guilds: moving beyond the pairwise perspective on mutualisms. *American Naturalist* 162: S10–S23.

Stappenbeck, T. S., Hooper, L. V., and Gordon, J. I. 2002. Developmental regulation of intestinal angiogenesis by indigenous microbes via Paneth cells. *Proceedings of the National Academy of Sciences USA* 99: 15451–15455.

Stat, M., Morris, E., and Gates, R. D. 2008. Functional diversity in coral-dinoflagellate symbiosis. *Proceedings of the National Academy of Sciences USA* 105: 9256–9261.

Steer, T. E., Johnson, E. T., Gee, J. M., and Gibson, G. R. 2003. Metabolism of the soyabean isoflavone glycoside genistein *in vitro* by human gut bacteria and the effect of prebiotics. *British Journal of Nutrition* 90: 635–642.

Stolyar, S., van Dien, S., Hillesland, K., Pinel, N., Lie, T. J., Leigh, J. A., and Stahl, D. A. 2007. Metabolic modelling of a mutualistic microbial community. *Molecular Systems Biology* 3: e92.

Stracke, S., Kistner, C., Yoshida, S., Mulder, L., Sato, S., Kaneko, T., Tabata, S., Sandal, N., Stougaard, J., Szczyglowski, K., and Parniske, M. 2008. A plant receptor-like kinase required for both bacterial and fungal symbiosis. Nature 417: 910–911

Strobel, G. 2006. Harnessing endophytes for industrial microbiology. *Current Opinion in Microbiology* 9: 240–244.

Strobel, G., and Daisy, B. 2003. Bioprospecting for microbial endophytes and their natural products. *Microbiology and Molecular Biology Reviews* 67: 491–502.

Suh, S-O., Noda, H., and Blackwell, M. 2001. Insect symbiosis: derivation of yeast-like endosymbionts within an entomopathogenic filamentous lineage. *Molecular Biology and Evolution* 18: 995–1000.

Tagg, J. R., and Dierksen, K. P. 2003. Bacterial replacement therapy: adapting "germ warfare" to infection prevention. *Trends in Biotechnology* 21: 217–223.

Tanaka, A., Christensen, M. J., Takemoto, D., Park, P., and Scott, B. 2006a. Reactive oxygen species play a role in regulating a fungus-perennial ryegrass mutualistic interaction. *The Plant Cell* 18: 1052–1066.

Tanaka, Y., Miyajima, T., Koike, I., Hayashibara, T., and Ogawa H. 2006b. Translocation and conservation of organic nitrogen within the coral-zooxanthella symbiotic system of *Acropora pulchra*, as demonstrated by dual isotope-labeling techniques. *Journal of Experimental Marine Biology and Ecology* 336: 110–119.

Taylor, D. L., and Bruns, T. D. 1997. Independent, specialized invasions of the ectomycorrhizal mutualism by two non-photosynthetic orchids. *Proceedings of the National Academy of Sciences USA* 94: 4510–4515.

Taylor, P. D., Day, T., and Wild, G. 2007. Evolution of cooperation in a finite homogenous graph. *Nature* 447: 469–472.

Taylor, T. N., Remy, W., Hass, H., and Kerp, H. 1995. Fossil arbuscular mycorrhizae from the early Devonian. *Mycologia* 87: 560–573.

Tchernov, D., Gorbunov, M. Y., Vargas, C. D., Yadav, S. N., Milligan, A. J., Haggblom, M., and Falkowski, P. G. 2004. Membrane lipids of symbiotic algae are diagnostic of sensitivity to thermal bleaching in corals. *Proceedings of the National Academy of Sciences USA* 101: 13531–13535.

Thimm, T., Hoffmann, A., Fritz, I., and Tebbe, C. C. 2001. Contribution of the earthworm Lumbricus rubellus (Annelida, Oligochaeta) to the establishment of plasmids in soil bacterial communities. *Microbial Ecology* 41: 341–351.

Thomas, G. H., Zucker, J., Sorokin, A., Goryanin, I., and Douglas, A. E. 2009. A fragile metabolic network adapted for cooperation in the symbiotic bacterium *Buchnera aphidicola*. *BMC Systems Biology* 3: 24.

Thompson, J. N. 2005. *The Geographical Mosaic of Coevolution*. University of Chicago Press, Chicago.

Thompson, J. N., and Cunningham, B. M. 2002. Geographic structure and dynamics of coevolutionary selection. *Nature* 417: 735–738.

Tielens, A.G.M., Rotte, C., van Hellemond, J. J., and Martin, W. 2002. Mitochondria as we don't know them. *Trends in Biochemical Sciences* 27: 564–572.

Timmins, J. N., Ayliffe, M. A., Huang, C. Y., Martin, W. 2004. Endosymbiotic gene transfer: Organelle genomes forge eukaryotic chromosomes. *Nature Reviews Genetics* 5: 123–126.

Toh, H., Weiss, B. L., Perkin, S.A.H., Yamashita, A., Oshima, K., Hattori, M., and Aksoy, S. 2006. Massive genome erosion and functional adaptations provide insights into the symbiotic lifestyle of *Sodalis glossinidius* in the tsetse host. *Genome Research* 16: 149–156.

Tovar, J., León-Avila, G., Sánchez, L. B., Sutak, R., Tachezy, J., van der Giezen, M., Hernández, M., Müller, M., and Lucocq, J. M. 2003. Mitochondrial remnant organelles of *Giardia* function in iron-sulphur protein maturation. *Nature* 426: 172–176.

Traulsen, A., and Novak, M. A. 2006. Evolution of cooperation by multilevel selection. *Proceedings of the National Academy of Sciences USA* 103: 10952–10955.

Treseder, K. K., and Cross, A. 2006. Global distributions of arbuscular mycorrhizal fungi. *Ecosystems* 9: 305–316.

Treseder, K. K., and Turner, K. M. 2007. Glomalin in ecosystems. *Soil Science* 71: 1257–1266.

Turelli, M. 1994. Evolution of incompatibility-inducing microbes and their hosts. *Evolution* 48: 1500–1513.

Turnbaugh, P. J., Ley, R. E., Mahowald, M. A., Magrini, V., Mardis, E. R., and Gordon, J. I. 2006. An obesity-associated gut microbiome with increased capacity for energy harvest. *Nature* 444: 1027–1031.

van Baalen, M., and Sabelis, M. W. 1995. The dynamics of multiple infection and the evolution of virulence. *American Naturalist* 146: 881–910.

van de Guchte, M., and 20 coauthors. 2006. The complete genome sequence of *Lactobacillus bulgaricus* reveals extensive and ongoing reductive evolution. *Proceedings of the National Academy of Sciences USA* 103: 9274–9279.

van den Giezen, M., Kiel, J., Sjollema, K. A., and Prins, R. A. 1998. The hydrogenosomal malic enzyme from the anaerobic fungus *Neocallimastix frontalis* is targeted to mitochondria of the methylotrophic yeast *Hansenula polymorpha*. *Current Genetics* 33: 131–135.

van der Heijden, M.G.A., Klironomos, J. N., Ursic, M., Moutoglis, P., Streitwolf-Engel, R., Boller, T., Wiemken, A., and Sanders, I. R. 1998. Mycorhizal fungal diversity determines plant biodiversity, ecosystem variability and productivity. *Nature* 396: 69–72.

van Dover, C. 2000. *The Ecology of the Deep-Sea Hydrothermal Vents*. Princeton University Press, Princeton, NJ.

van Rhijn, P., Fujishige, N. A., Lim, P. O., and Hirsch, A. M. 2001. Sugar-binding activity of pea lectin enhances heterologous infection of transgenic alfalfa plants by *Rhizobium leguminosarum* biovar *viciae*. *Plant Physiology* 126: 133–144.

Vandamme, P., Goris, J., Chen, W-M., de Vos, P., and Willems, A. 2002. *Burkholderia tuberum* sp. nov. and *Burkholderia phymatum* sp. nov., nodulate the roots of tropical legumes. *Systematic and Applied Microbiology* 25: 507–512.

Vandenkoornhuyse, P., Ridgway, K. P., Watson, I. J., Fitter, A. H., and Young, J.P.W. 2001. Co-existing grass species have distinctive arbuscular mycorrhizal communities. *Molecular Ecology* 12: 3085–3095.

Vermeij, G. J. 2004. *Nature: An Economic History*. Princeton University Press, Princeton, NJ.

Völkl, W. 1992. Aphids or their parasitoids—who actually benefits from ant-attendance. *Journal of Animal Ecology* 61: 273–281.

von Dohlen, C. D., Kohler, S., Alsop, S. T., and McManus, W. R. 2001. Mealybug beta-proteobacterial endosymbionts contain gamma-proteobacterial symbionts. *Nature* 412: 433–436.

von Heijne, G. 1986. Why mitochondria need a genome. *FEBS Letters* 198: 1–4.

Vrieling, K., Smit, W., and van der Meijden, E. 1991. Tritrophic interactions between aphids *Aphis jacobaea* Schrank), ant species, *Tyria jacobaeae* L., and

Senecio jacobaea L. lead to maintenance of genetic variation in pyrrolizidine alkaloid concentration. *Oecologia* 86: 177–182.

Walter, R. F., and McFadden, G. I. 2005. The apicoplast: a review of the derived plastid of apicomplexan parasites. *Current Issues in Molecular Biology* 7: 57–79.

Warner, M. E., Fitt, W. K., and Schmidt, G. W. 1999. Damage to photosystem II in symbiotic dinoflagellates: A determinant of coral bleaching. *Proceedings of the National Academy of Sciences USA* 96: 8007–8012.

Way, M. J. 1963. Mutualism between ants and honeydew-producing Homoptera. *Annual Review of Entomology* 8: 307–344.

Wedin, M., Doring, H., and Gilenstam, G. 2004. Saprotrophy and lichenization as options for the same fungal species on different substrata: environmental plasticity and fungal lifestyles in the *Stictis-Conotrema* complex. *New Phytologist* 164: 459–465.

Weeks, A. R., Turelli, M., Harcombe, W. R., Teynolds, K. T., and Hoffmann, A. A., 2007. From parasite to mutualist: rapid evolution of *Wolbachia* in natural populations of *Drosophila*. *PLoS Biology* 5: e114.

Weerasinghe, R. R., Bird, D. M., and Allen, N. S. 2005. Root-knot nematodes and bacterial Nod factors elicit common signal transduction events in *Lotus japonicus*. *Proceedings of the National Academy of Sciences USA* 102: 3147–3152.

Welburn, S. C., and Maudlin, I. 1999. Tsetse-trypanosome interactions: rites of passage. *Parasitology Today* 15: 399–403.

Westerkamp, C., and Gottsberger, G. 2000. Diversity pays in crop pollination. *Crop Science* 40: 1209–1222.

Weis, V. M., Davy, S. K., Hoegh-Guldberg, O., Rodriguez-Lanetty, M., and Pringle, J. R. 2008. Cell biology in model systems as the key to understanding corals *Trends in Ecology and Evolution* 23: 369–376.

Whitehead, L. M., and Douglas, A. E. 1993. Populations of symbiotic bacteria in the parthenogenetic pea aphid (*Acyrthosiphon pisum*) symbiosis. *Proceedings of the Royal Society of London B* 254: 29–32.

Wickler, W. 1966. Mimicry in tropical fishes. *Philosophical Transactions of the Royal Society of London B* 251: 473–474.

Wiener, P. 2000. Antibiotic production in a spatially structured environment. *Ecology Letters* 3: 122–130.

Wijffels, R. H. 2007. Potential of sponges and microalgae for marine biotechnology. *Trends in Biotechnology* 26: 467–471.

Wilding, N., Collins, N. M., Hammond, P. M., and Webber, J. F. 1989. *Insect-Fungus Interactions*. Academic Press, London.

Wilkinson, D. M., and Sherratt, T. N. 1999. Horizontally acquired mutualisms, an unsolved problem in ecology? *Oikos* 92: 377–384.

Wilkinson, T. L., Adams, D., Minto, L. B., and Douglas, A. E. 2001. The impact of host plant on the abundance and function of symbiotic bacteria in an aphid. *Journal of Experimental Biology* 204: 3027–3038.

Wilson, E. O. 1971. *The Insect Societies*. Harvard University Press, Cambridge, MA.

Wilson, E. O., and Hölldobler, B. 2005. The rise of ants: a phylogenetic and ecological explanation. *Proceedings of the National Academy of Sciences USA* 102: 7411–7414.

Wilson, M. 2005. *Microbial Inhabitants of Humans*. Cambridge University Press, Cambridge.

Wood, R. 1997. *Reef Evolution*. Cambridge University Press, Cambridge.

———. 1998. The ecological evolution of reefs. *Annual Review of Ecology and Systematics* 29: 179–206.

Woyke, T., and 17 coauthors. 2006. Symbiosis insights through metagenomic analysis of a microbial consortium. *Nature* 443: 950–952.

Wu, M., and 28 coauthors. 2004. Phylogenomics of the reproductive parasite *Wolbachia pipientis* wMel: A streamlined genome overrun by mobile genetic elements. *PLoS Biology* 2: 327–341.

Xavier, R. J., and Podolsky, D. K. 2007. Unravelling the pathogenesis of inflammatory bowel disease. *Nature* 448: 427–434.

Xu, J., and 18 coauthors. 2007. Evolution of symbiotic bacteria in the distal human intestine. *PLoS Biology* 5: e156.

Yao, I., and Akimoto, S-I. 2002. Flexibility in the composition and concentration of amino acids in honeydew of the drepanosiphid aphid *Tuberculatus quercicola*. *Ecological Entomology* 27: 745–752.

Yoshida, N., Oeda, K., Watanabe, E., Mikami, T., Fukita, Y., Nishimura, K., Komai, K., and Matsuda, K. 2001. Chaparonin turned insect toxin. *Nature* 411: 44.

Young, J. M., Kuykendall, L. D., Martinez-Romero, E., Kerr, A., and Sawada, H. A. 2001. Revision of *Rhizobium* Frank 1889, with an emended description of the genus, and the inclusion of all species of *Agrobacterium* Conn 1942 and *Allorhizobium undicola* de Lajudie et al. 1998 as new combinations: *Rhizobium radiobacter*, *R. rhizogenes*, *R. rubi*, *R. undicola* and *R. vitis*. *International Journal of Systematic and Evolutionary Microbiology* 51: 89–103.

Yu, D. W., and Pierce, N.E. 1998. Castration parasite of an ant-plant mutualism. *Proceedings of the Royal Society of London B* 265: 375–382.

Yuan, X., Xiao, S., and Taylor, T. N. 2005. Lichen-like symbiosis 600 million years ago. *Science* 308: 1017–1020.

Zabalou, S., Riegler, M., Theodorakopoulou, M., Stauffer, C., Savakis, C., and Bourtzis, K. 2004. *Wolbachia*-induced cytoplasmic incompatibility as a means for insect pest population control. *Proceedings of the National Academy of Sciences USA* 101: 15042–15045.

Zabinski, C. A., Quinn, L. and Callaway, R. M. 2002. Phosphorus uptake, not carbon transfer, explains arbuscular mycorrhizal enhancement of *Centaurea maculosa* in the presence of native grassland species. *Functional Ecology* 16: 758–765.

Zaidman-Rémy, A., Herve, M., Poidevin, M., Pili-Floury, S., Kim, M. S., Blanot, D., Oh, B. H., Ueda, R., Mengin-Lecreulx, D., and Lemaitre, B. 2006. The *Drosophila* amidase PGRP-LB modulates the immune response to bacterial infection. *Immunity* 24: 463–473.

Zientz, E., Dandekar, T., and Gross, R. 2004. Metabolic interdependence of obligate intracellular bacteria and their insect hosts. *Microbiology and Molecular Biology Reviews* 68: 745–770.

Zyrek, A. A., Cichon, C., Helms, S., Enders, C., Sonnenborn, U., and Schmidt, M. A. 2007. Molecular mechanisms underlying the probiotic effects of *Escherichia coli* Nissle 1917 involve ZO-2 and PKCζ redistribution resulting in tight junctions and epithelial barrier repair. *Cellular Microbiology* 9: 804–816.

Index